The
Energy
Plan

Leabharlanna Poiblí Chathair Bhaile Átha Cliath
Dublin City Public Libraries

Marino Branch
Brainse Marino
Tel: 8336297

Comhairle Cathrach
Bhaile Átha Cliath
Dublin City Council

Due Date	Due Date	Due Date

(2018 World Cup winners) and UEFA, leading a team of scientists to create guidelines for nutrition in world football.

He was previously elected President of the Royal Society of Medicine's Food and Health Forum and he has his own private practice in Harley Street, where he sees performing artists, business executives and clients from all walks of life, helping them achieve their physical and mental best.

After completing his MSc in Sport & Exercise Nutrition at Loughborough University and International Olympic Committee (IOC) Diploma, he was recruited to join The English Institute of Sport in 2006. He writes a monthly column in *The Telegraph* and is a technical consultant for the BBC's *Good Food* brand. He is a registered Sport and Exercise Nutritionist (SENr).

The
Energy
Plan

James Collins

Vermilion
LONDON

1 3 5 7 9 10 8 6 4 2

Vermilion, an imprint of Ebury Publishing,
20 Vauxhall Bridge Road,
London SW1V 2SA

Vermilion is part of the Penguin Random House group of companies
whose addresses can be found at global.penguinrandomhouse.com

Penguin
Random House
UK

First published by Vermilion in 2019

www.penguin.co.uk

A CIP catalogue record for this book is available
from the British Library

ISBN 9781785042294

Typeset in 10.5/14 pt Sabon LT Std
by Integra Software Services Pvt. Ltd, Pondicherry

Printed and bound in Great Britain by Clays Ltd, Elcograf S.p.A.

Penguin Random House is committed to a sustainable future for our business, our readers and our planet. This book is made from Forest Stewardship Council® certified paper.

The information in this book has been compiled by way of general guidance in relation to the specific subjects addressed, but is not a substitute and not to be relied on for medical, healthcare, pharmaceutical or other professional advice on specific circumstances and in specific locations. So far as the author is aware the information given is correct and up to date as at January 2019. Practice, laws and regulations all change, and the reader should obtain up to date professional advice on any such issues. The author and publishers disclaim, as far as the law allows, any liability arising directly or indirectly from the use, or misuse, of the information contained in this book.

For my family, friends and everyone who believes in the
power of nutrition

Contents

Introduction: Achieving Your Personal Best

When I first started working as a nutritionist in elite sport over a decade ago, the most common thing I would hear from athletes was: 'James, I'm here because Coach says I need to lose weight.'

Fortunately things have moved on just a little since then.

Nutrition has become an integral part of every professional sports organisation, an area of substantial investment for the leading teams. We work within the sports medicine and science teams to produce nutrition plans that give the athletes the best chance of success when it matters most. We help them fuel their bodies in harmony with their training and competition schedules, boost their energy levels, recover effectively from regularly putting their bodies on the line and maintain a robust immune system.

And outside of sport, 'weight loss' (or, as we should say, reducing body fat) might be the reason many of us first start looking at our nutrition, but there's much more to it than that now. As a nation we are more engaged with the nutrition advice and information that fills the pages of newspapers, lifestyle websites and social-media feeds than ever before. But the pressure to constantly look good means that there is often a trade-off with energy levels, mood and productivity each day.

I've seen first-hand examples in weight-category sports such as horse racing, boxing and judo, where the performers' focus is on hitting the number on the scales in order to compete, but in doing so they have no energy left to do themselves justice. It's the same principle for anyone who has been on a diet and found themselves hungry and fatigued, with mood and motivation levels plummeting:

It's no good looking great if you don't have the energy to deliver a performance.

The Energy Plan is an approach to peak performance that I have developed over more than ten years of working in sport as a nutritionist, with some of the best athletes and football players in the world, and in my private practice on Harley Street, where my clients come from all walks of life – musicians, dancers, artists, actors, entrepreneurs and professionals of any age. The same principles apply to performers in any discipline.

Performance can seem like a hard and scary word, but ultimately helping you use nutrition to perform at your personal best in all aspects of your life is what this book is about.

My journey began when I worked with the British Athletics team prior to the Beijing Olympic Games in 2008, and continued as part of an approach to nutrition that helped power Team GB to success at the London 2012 Olympic Games.

My use of food to improve performance led Arsène Wenger to recruit me for Arsenal Football Club in 2010 as the club's first nutritionist. I spent seven seasons working closely with world-class players such as Alexis Sánchez, Mesut Özil, Alex Oxlade-Chamberlain and Héctor Bellerín, tailoring their nutritional needs to meet the demands of packed fixture lists, late-night games, travel to some of the further reaches of Europe for Champions League away games and to give them the best chance of recovery and to minimise injury and illness.

I have also worked with the England football team, and, beyond these shores, I was a consultant to the France Football Federation for their 2018 World Cup-winning campaign, and

UEFA, leading a team of scientists to create guidelines for nutrition in world football.

My methods are practical and straightforward to apply, and my results deal in the only real currency all sportsmen and women are interested in: improving performance on the field of play. This may mean just making a few functional changes to meals and timings, rather than turning your eating habits upside down and trying to create an unsustainable plan.

Outside of sport, I have worked with music and dance organisations in London to help them meet the gruelling demands of rehearsals and performances night after night in the West End or on the road.

I was elected president of the Royal Society of Medicine's Food & Health Forum, and I write regular columns for *The Telegraph* newspaper and the BBC. I was part of the 2018 Sport Relief project in which I helped entertainers who were far from being athletes find the energy and confidence that was missing from their lives, and discover that it's never too late to improve your health and quality of life.

The Energy Plan is as applicable to you or me as it is to sports stars. At its heart is the simple premise that **food is fuel**. Our body and mind run on this resource, and it is only through fuelling in a targeted and deliberate manner that we can look and feel the way we want, and enjoy the benefits of a bountiful reservoir of energy.

There is a lot of noise around nutrition, a lot of conflicting advice and polar-opposite dietary plans that simply leave many of us in a state of confusion. Many of the plans and programmes we're sold have little in the way of evidence-based research to support them. So my approach – the approach we use throughout sport – is to apply the science to make the biggest possible impact on your life. I'm going to let you in on the secrets of how the best performers manage their energy to deliver on the biggest stages.

The Energy Plan is about more than just being lean (though it can certainly help you achieve this). It isn't an unsustainable fad diet that you will give up on in a matter of weeks. It isn't a rigid, unchanging menu that will see you eat the same sort of meals every day.

It's a feasible guide to using nutrition for the *rest of your life*, using your fuel to meet your body's daily demands, because those demands change every day, week, month and year of your life. If you're pushing your body to its limits at the gym one day and then spending the next two sitting at your desk, your body will demand a different approach from your nutrition. Eating the same thing each day as part of a 'straightjacket' diet plan is simply not going to cut it.

With **The Energy Plan** I will show you, among many other things, how to:

- Apply the scientific secrets the world's top athletes use to fuel their bodies for success.
- Address your energy levels so you can enjoy sustained peaks and reduced troughs and be a top performer when it matters most.
- Use the full range of nutrients – including carbs – at the right times to give you energy tailored to your needs.
- Reduce your body fat and maintain your muscle mass without compromising on your energy levels.
- Use nutrition and exercise to keep you strong and less likely to get ill.
- Fuel up and recover effectively for a major physical challenge in your life, whether that's your first 5k run or the latest in a long line of endurance events.
- Use food from meals and snacks to power your day – without being overly reliant on caffeine.
- Get everything from your food no matter what your dietary requirements are – committed carnivore, vegetarian or vegan, and everything in between.
- Build a plate of food like the top performers do.
- Enjoy a full and active social life while still achieving your goals.

- Shop! With a targeted approach to your groceries you can have all the building blocks for quick-to-make, healthy and delicious meals at the ready.
- Use supplements more strategically – you might think you can't have too much of a good thing, but those daily vitamin tablets might be a waste of your money and may even jeopardise your health.
- Recharge and sleep more effectively to recover from the demands of your daily life.
- Manage your nutrition on the road – from international travel for work to a family holiday.
- Age anything but gracefully by adapting your habits to ensure that you feel like someone much younger as mid- and later-life approach.

I want to cut through the noise around nutrition and offer you **The Energy Plan**, a clear, scientific approach that might upset some people in the billion-dollar diet and supplement industries, but which will give you the best chance of looking good, feeling great, avoiding injury and illness and having enough energy to put in a gold-medal performance in your own life.

How to Use This Book

The Energy Plan is structured into three parts. In **Part I: The Energy Balance**, we will look at the core principles behind the Energy Plan: energy's journey through your body and your metabolism (your **engine**); the nutrients that make up the **fuel** for your engine; and the needs you fulfil when you step on **the accelerator** and move your body. While I do recommend that you read through the whole book chronologically to get the most from it, if you feel you're already sufficiently well-versed in the role of nutrients, your metabolism and physical activity, then you can move straight on to the plan itself in Part II.

Part II: Your Energy Plan is the practical application section, in which you will develop your own Energy Plan. You will first establish your goals, and then look at how you can build your own **performance plates** of food, with the relevant amounts of carbohydrate, protein, fat and micronutrients depending on your needs for that day, like the elite athletes I work with. We will look at building these plates into your day and then your week, so you have the skills to manage your nutrition all of the time, and then we will look at some of the challenges to your Energy Plan and what you can do if the wheels come off. We'll address the main challenges posed by your workplace and other environments, so you can enjoy **sustained peaks and reduced troughs** throughout your working day, and how to relax things at the weekend. And finally, we'll look at your shopping habits and the preparation you can do to make your Energy Plan fit into your lifestyle as well as possible.

In **Part III: Sustainable Energy**, we'll address the big issues your Energy Plan will involve. We'll look at your **sleep**, and how to manage this most important opportunity in your day to recharge; you'll learn how to stay strong and healthy thanks to robust **immunity**, particularly during the colder months; I'll share the secrets to ensuring **travel** – whether long- or short-haul – doesn't derail your performance; we'll lift the lid on the Wild West that is food **supplements**, and put in place a process to decide whether you need to take any; and, to finish, we'll look at the one thing none of us can fight – time – and show you how your Energy Plan can help you to manage the effects of the **ageing** process. And if you think that chapter's not for you, remember that the effects of ageing begin to make themselves felt in your thirties.

Your Home Essentials

This isn't an endless list of expensive kit – these are just some of the essentials to keep you on track when executing your Energy Plan. You might already own much of this, or think some things (such as the coffee machine or tableware) aren't essential, so please feel free to cherry-pick those that are essential to you and your Energy Plan. It's worth returning to this list after you've read to the end of the book, as your thoughts on what you need may change.

- Trainers
- Gym membership or home weights
- Comfortable workout kit (it's important to look and feel the part)

- Hands-free phone kit (for work calls on the move, to increase your activity levels)
- Earplugs and eye mask (for travel – see Chapter 13)
- Water bottle (if you prefer flavoured water, choose one that has a fruit infuser)
- Protein shaker (to mix protein powder with liquid smoothly, leaving no lumps)
- Notebook (for your regular check-in – see Chapter 8)
- A gym bag
- Coffee capsule machine (to understand your caffeine dose and timing, and keep it consistent)
- Storage boxes (for lunch and snacks at work)
- Kitchen essentials: sharp knives, chopping boards, non-stick pans, griddle, saucepans, baking trays
- Heavy-set glasses and cutlery (see Chapter 7, Winning Behaviours)

PART I

The Energy Balance

CHAPTER 1

The Engine

Your breakfast might be a couple of poached eggs, golden yolks spilling slowly on to rich green avocado, freckled with chilli and mint, all sitting atop a slice of sourdough bread, in your favourite café – deliciously Instagrammable. It might be a humble bowl of cornflakes at your kitchen table, doused in semi-skimmed milk and awaiting that first reassuringly crunchy bite. Or it could simply be a slice of toast in your hand as you race out the door to catch the train. You might even choose to skip breakfast altogether, going for a run first thing and then hopefully grabbing something on the way in to work.

But no matter what it is, and no matter how aware of it you are, this is how your engagement with energy for the day begins. Whether you're conscious of it or not, you are already making decisions that will have an impact on your energy in both the near future – later this morning, afternoon or evening – and the longer term – the habits that will continue tomorrow, the impact on your body in the coming days, weeks, months and years.

Performance nutrition is about making strategies to improve your energy levels – whether that's at work, in your fitness regime or at home – and instilling the know-how and habits so that you can sustain your progress. Part of my role as a sports nutritionist is to educate, and while some of my clients – whether elite athletes or active businessmen and women – come to me with a good working knowledge of their body and how it uses food to produce energy, for many others biology lessons from school seem like a long time ago and their understanding might not be so clear.

Too much nutritional advice today tries to dazzle, with expressions like 'metabolic', 'antioxidants' and 'phytonutrients', but without a basic grasp of how our bodies work, we can't fully understand how nutrients are used as fuel, meaning that each new piece of nutrition information we hear or read just layers on more confusion.

That's why I like to start by getting back to basics: looking under the bonnet of these high-performance vehicles we all walk around in every day.

The Energy Journey

Whether it's eggs and avocado, cornflakes or a slice of toast, your body, unlike your Instagram followers, won't discriminate: they're all sources of energy that will propel your body and mind through your day. And, as you bite into that initial mouthful of food, energy's journey begins in earnest.

As you chew you are increasing the surface area of food for the benefit of digestive chemicals called **enzymes**, which break food down into particles small enough to be used by the body. Some of the key **micronutrients** (vitamins and minerals, which we'll discuss in more detail in Chapter 2) from fruit and vegetables, such as dietary nitrates from beetroot, begin a crucial part of their digestion in the mouth.

The mouth also contains receptors that link to the brain's pleasure and reward centres. Research has shown that endurance athletes get a performance boost simply from using carbohydrate as a mouthwash and spitting it out, without needing to swallow and digest in the traditional manner – about as quick an energy shortcut as you can imagine.[1]

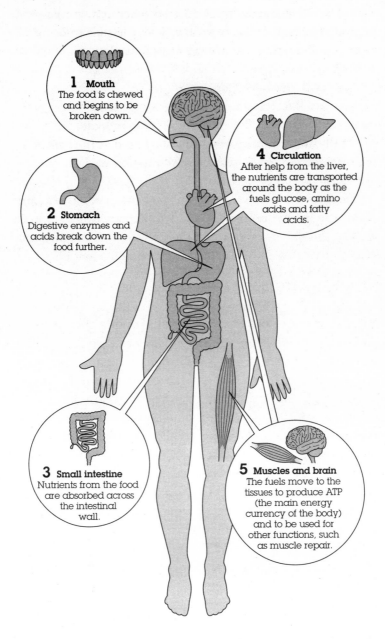

1 Mouth
The food is chewed and begins to be broken down.

4 Circulation
After help from the liver, the nutrients are transported around the body as the fuels glucose, amino acids and fatty acids.

2 Stomach
Digestive enzymes and acids break down the food further.

3 Small intestine
Nutrients from the food are absorbed across the intestinal wall.

5 Muscles and brain
The fuels move to the tissues to produce ATP (the main energy currency of the body) and to be used for other functions, such as muscle repair.

The energy journey

When you swallow your food, it travels down the oesophagus and into the stomach, which releases more digestive enzymes and acid to help break down the food, and to sterilise it all. The stomach acts as a short-term store and gradually passes the broken-down food to the intestines, where the digestive process will be completed.

The small intestine – it's actually about six or seven metres long, depending on the individual – with some help from the liver, finishes breaking down the food so that it's now made up of molecules ready for the body to use. These digested food molecules then pass through the wall of the small intestine to the liver for further processing and into the bloodstream. Once the small intestine has finished this process, what is left passes through the large intestine, which absorbs the water and electrolytes – and I'm sure you're familiar with how this particular story ends... when you sit down on the toilet.

This system, one long, uninterrupted line that begins with your mouth and ends with your large intestine, is called the **gastrointestinal (GI) tract.** We will discuss your gut (or GI) health later in the book, when we talk about immunity.

In the meantime, however, your energy, which started life as your breakfast, is now in your blood, and its journey is about to get a lot more action-packed.

Your High-Performance Vehicle

Performance nutrition has been adopted in Formula 1 motor racing in recent years. I must confess that I'm not much of a petrolhead but the motor car does offer a good representation of the human body and our relationship with food.

The process I've just described is **digestion**, in which we take food and break it down into smaller parts, first with our teeth and then with our small intestine, so that it exists in a currency our body can use. The Energy Plan approach is that **food is fuel**, and the process of eating and digestion is like fuelling our engine at the petrol station. Our metabolism, which is the next state of this process, is the car engine. It takes its fuel, in the form of digested food, and turns it into energy to support the body's maintenance – the growth and repair of organs and cells – and movement. Your walk to work, your morning HIIT class, even your immune health – they're all thanks to your metabolism. Making sure that you have taken on enough fuel for your bodily functions and your day's activities, so that you're not underfuelled and in energy deficit, or overfuelled and with an energy surplus, is the key to achieving an energy balance. You don't want the fuel gauge to be reading 'empty' by 4pm, just as you don't want to feel like you've packed the trunk too full through constant overfuelling.

With your fuel now in the bloodstream thanks to the digestive process, it can circulate in the body's useable currency: protein broken down into amino acids, carbohydrates in the form of glucose, and fat as fatty acids, all of which we'll discuss in more detail in the next chapter. Water, vitamins and minerals do not need breaking down in the same manner as their molecules are already of a suitable size for the intestines to absorb.

Your cells take up this fuel and use it to create adenosine triphosphate (ATP), which is the main energy currency of the body and the only one organs like your brain really trade in. ATP is the currency of choice for your muscles and, in essence, for the Energy Plan.

Our body contains different types of muscle: skeletal muscle, which powers voluntary movements like walking or picking something up; cardiac muscle to pump blood from the heart; and smooth muscle, which coordinates the involuntary movements of organs like the stomach and intestines. All of these types of muscle require an expenditure of energy.

Within all the types of muscle cells there are a host of power generators called mitochondria. These take fuel and convert it into ATP, which powers your body's needs:

movement, cell growth and repair and cognitive function. We'll return to these power generators later in the book, in Chapter 2, as they are something we can exercise some control over with our Energy Plan; they can have a positive effect on goals like losing body fat.

Your skeletal muscles contain some intricate architecture to help you move, just as a car contains the requisite man-made mechanics to power its movement. In the body's case, this involves using ATP to power the muscles to contract, which creates movement.

The Three Avenues of Energy Expenditure

1. **Resting metabolic rate (RMR):** the energy the body requires for normal functioning at rest and the largest component, accounting for 60 to 75 per cent of energy expenditure. The more lean muscle mass you have, the higher your resting metabolic rate, which is why women tend to have a lower rate than men; and it decreases with age, by around 2 or 3 per cent per decade. So that's between two-thirds and three-quarters of your energy output that takes care of itself – assuming you're prepared to put the building blocks in place.

2. **Thermic effect of food (TEF):** the energy your body uses to digest and process food, accounting for around 10 per cent of your daily energy expenditure. This depends on the kind of food you eat: processing carbohydrates and fat uses between 5 and 15 per cent of the consumed energy, but protein uses a whopping 25 to 30 per cent.

3. **Physical activity:** refers to everything from involuntary movements like shivering and fidgeting to your morning cycle to work. It is the most effective way to increase energy output and, fairly unsurprisingly, where we see the biggest variation: a sedentary person can expend as little as 100 kcal in a day, while an endurance athlete might hammer out over 6,000 kcal.

Currency Exchange

We will reference energy's journey and the various important aspects of our body, such as the mitochondria, throughout the book, so it might be useful to refer back to this chapter later. What is important to grasp now is the idea that **your body is an engine**, and it uses food as fuel to power all of its functions, from cellular activity through to movement.

With the athletes I work with, we don't talk about being on 'diets' or anything that suggests a short-term fix before returning to a status quo. Instead the athletes are on constantly evolving Energy Plans, which are flexible and can be changed to meet the demands of the day, or a longer period of time, such as a week or a season. They might run an energy deficit for a while to achieve a goal, but they will then adjust to another form of Energy Plan. There is never a sense of being 'finished'; the Energy Plan is a sustainable way of life.

It's important for me to emphasise that I want you to think about food in a **positive** way. No matter what you eat, whether it's the picture-perfect poached eggs and avocado or the bowl of cornflakes, it all makes the same journey we've just outlined, even if its constituent parts might have very different effects. Food is fuel, of course, but it also has the capacity to be elegant, enticing and aspirational; it can be warming, reassuring and comforting; and it brings people together as the centrepiece of some of the biggest and most important moments of our lives, such as weddings, birthday parties and first dates – or as a takeaway in front of the television with your friends or loved ones.

I want to be clear now that the Energy Plan isn't about depriving you of the food and drink you love. It's about giving you the tools to understand your intake and be able to **tailor your food to meet your goals**. Better nutrition starts by *increasing* the variety of foods in your diet, not restricting, to deliver the full spectrum of nutrients each day. In my own Energy Plan I still find room to enjoy a burger. If I meet a client in a café, I have a coffee (flat white, please – 'extra hot' is my only higher-maintenance request). And I also think, if you're going to put so much dedication into living well for the majority of the time, you deserve to treat yourself from time to time.

But it is important to be aware of the constituent parts of the body's fuel, so that you know exactly what you're 'filling up' with and how this will affect you throughout your day. Our body is both an engine and a very clever currency exchange system, which takes the parts of food and breaks them down into different currencies, several times over, to deliver our body's needs. So, now that we have a sense of energy's journey, let's take a look at the fuels we have at our disposal to best keep our engine ticking over.

CHAPTER 2

The Fuels

An international footballer might complete 300 training sessions each season, but they will also eat 1,050 meals. That's 1,050 opportunities to fuel and adapt to their training. It's important for athletes to understand the power of nutrition to fuel their bodies; and make no mistake – it's the same for you. Each meal has the power to positively or negatively contribute towards reaching your goals.

Looking at nutrition through this lens is an ability that you can develop, ensuring you can still enjoy your food while also better understanding the building blocks of which it is composed. You will be able to see the food on your plate in relation to how it's helping you achieve your goals for that day and beyond, and learn that, when it comes to the nutrients that make up your fuel, there are by definition no nutrients that are inherently bad or good. It's all a question of context.

The Right Fuel for the Engine

For our engine to function properly, our fuel must include the essential nutrients. A nutrient is quite simply a substance that is essential for the maintenance of life and growth. Alongside water, there are two types of nutrients: **macronutrients**, which include carbohydrates, fats and protein and are required in large doses, and **micronutrients**, which are made of vitamins and minerals and are, as the name suggests, required by the body in smaller quantities.

Key Fuels

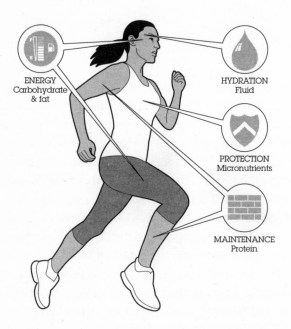

This simple labelling of your fuel's composition doesn't tell the whole story, though, as there is a lot of overlap in the roles of the nutrients (protein, for example, can also be used as fuel for energy production in times of need; 'healthy' fats – see page 31 – as protection), but it is a sufficient working model for understanding the function of each food, and we will return to this model throughout the book. **Energy, maintenance, protection** and **hydration** are fuel's key roles, so let's take a look at the part each nutrient plays in the Energy Plan.

Carbohydrates: The Rapid Energy Source

Carbohydrates are one of the main fuels for the central nervous system and the muscles. The body has limited stores of carbohydrate (in comparison to its large reservoir of fat) and carbohydrate is stored in both the liver and the muscles. Depending on the size of the person, approximately 80–110g is stored in the liver, and 300–600g in the muscles. It is quickly exhausted by vigorous exercise.

The liver helps to maintain our blood sugar (glucose) levels, which are tightly controlled to prevent hypoglycaemia (low blood sugar). I'm sure we're all familiar with that feeling – it causes dizziness, fatigue and sweaty palms.

The carbohydrate stored in the muscles (called in this context 'glycogen', as it is for that stored in the liver), is the fuel store for movement. Depleted muscle glycogen stores can cause muscle fatigue, especially during hard exercise. This has been termed 'hitting the wall' in the context of endurance sports, and is when the muscles run out of fuel, causing sudden fatigue, and the legs feel like they're turning to jelly.

Despite the mass of 'new' nutrition and exercise advice filling the media over the last couple of decades, how our bodies use fuel hasn't actually changed over the last 20 years. During low-intensity exercise, fat is the predominate fuel used by muscles. As this increases to moderate-intensity work, the muscles use a mixture of fat and carbohydrate. But at higher exercise intensities (such as sprinting, hard exercise classes and hill climbs), your engine will switch to using carbohydrate as its main fuel source, as it can be converted into energy much quicker than fat (which we will come on to next) (see diagram overleaf).

Our body's glycogen stores need regularly topping up with carbohydrates, and it is through understanding when the body needs carbohydrates that we can more easily plan when to include them in meals. It's handy to think of our carbohydrate stores as a fuel gauge: for hard work the levels

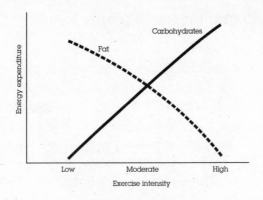

How our bodies use different fuels

need to be topped up; for certain training scenarios they can be low ('training low', which we'll talk about in Chapter 3). However, walking around with a chronic low intake can lead to fatigue and a host of physiological consequences, which we will discuss later in the chapter.

For the athletes I work with and for you, it's not about having a high- or a low-carb diet, it's about **matching your intake to your activity levels**. Not only that, but blood glucose (converted into ATP) is the main fuel used by the brain – and the brain is the most energy-hungry organ you have, demanding a whopping 20 per cent of the body's energy consumption[1] – so, if you'll pardon the pun, it's a no-brainer that you'll need some carbs in your diet.

There are various types of carbohydrates, with suitably scientific names like monosaccharides (such as glucose and fructose) and polysaccharides (such as starch), but for our purposes we can broadly categorise carbs into two, easier to remember labels. Step forward the glycaemic index.

The glycaemic index (GI) ranks foods depending on how quickly they are digested and broken down into blood glucose, on a scale of zero to 100, with pure glucose (sugar) having a value of 100.

It's important also to consider some factors that affect the GI of foods.

Think of carbohydrate stores as a fuel gauge

Cooking foods containing carbohydrates causes a breakdown in the starchy structure; for example, the longer pasta is cooked for, the higher its GI becomes. And cooling foods such as rice, pasta and potatoes after cooking, rather than eating them hot, lowers the GI.

The ripeness of a food increases its GI, such as a banana that's turned from green to yellow. More processed foods have smaller particles and a higher GI; so cornflakes are higher-GI than porridge oats. The amount of protein, fat and soluble fibre in a food lowers its GI.

High GI foods are more refined carbohydrates, which are digested very quickly and cause a sharp rise and similarly steep fall in blood glucose levels. Examples include white bread, cereal bars and puffed-rice cereals.

FREE SUGARS

Free sugars are simple sugars that have been added to food. And they're called 'free' because they're not inside the cells of food we eat – this is the crucial difference between whole fruit and fruit juice. It's important to be aware of free sugars and limit your intake to less than 30 g per day (six teaspoons). Be aware that these are always listed within 'total sugars' on the label, not separately. Common offenders include coffee syrups, smoothies and juices, breakfast bars, yoghurts… these sugars are often hidden in things you think are healthy choices. But as you will see in Part II, the Energy Plan will highlight better-quality alternatives.

Low GI foods are complex or starchy carbohydrates which are broken down in the body more slowly and give a slower, more sustained energy release. Low GI foods typically contain more fibre. Examples include rye bread, bulgur wheat and porridge oats. See the carbohydrates table in the appendix for the GI ranking of other carbohydrates.

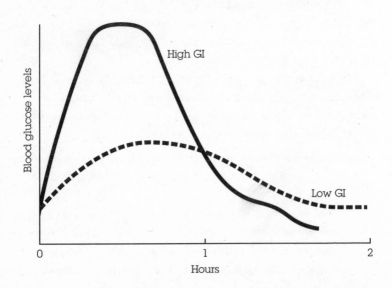

As this graph shows, the most desirable form of carbohydrates are those at the lower GI end of the spectrum, as they will provide a more sustained form of energy release into your blood. You are less likely to suffer a mid-morning crash and subsequent hunger after a bowl of porridge oats for breakfast than you are with a couple of slices of white toast. It's worth bearing in mind that, with each meal, we eat a mix of other macronutrients – such as protein and fat – which will have a hand in slowing digestion and the subsequent rise in blood glucose (known as reducing the glycaemic load of the meal). The quantity of carbohydrate is important too: a small (say, 10 g) bite of a high GI food, for example, will have less of an effect than a larger portion.

BLOOD GLUCOSE EXPLAINED

After eating a meal containing carbohydrate, blood glucose levels increase, which causes the hormone insulin to be released from the pancreas. This hormone signals to the cells to take up more glucose for energy or storage. When glucose levels fall another hormone, glucagon, sends the opposite signal to the liver, to increase blood glucose levels. Over time chronic issues can happen to the machinery - for example, Type 2 diabetes occurs when the cells in the muscles, body fat and liver stop responding to insulin (become insulin-resistant), leaving blood glucose levels high after eating. This is an increasing prob-lem and is covered in more depth in Chapter 15 on ageing.

The glycaemic index shouldn't necessarily be read as catego-rising foods as being more or less inherently 'healthy' overall; it just provides a useful way of classifying the types of fuel derived from different carbohydrates. As a general rule, most of the carbohydrates we eat each day should be low GI, fibrous options. Together with regular mealtimes, lower glycaemic index foods can play a role in regulating blood glucose and energy levels throughout the working day. But there will be occasions in a training setting when high GI foods are useful.

Fibre

Fibre describes plant-based carbohydrates that aren't absorbed by the body. It plays an important role in slowing carbohydrate absorption (measured by the glycaemic index) and blood glucose levels, and in improving satiety (in other words, feeling full) from meals. Fibre is important for a healthy digestive system, normalising bowel movements and feeding 'healthy' gut bacteria. Fibre is often classed as either soluble or insoluble, depending on its solubility in water (the main difference is that insoluble fibre isn't broken down as it passes through the digestive tract). High-fibre foods include: wholegrains (oats, barley, rye), beans, pulses, nuts, seeds and root vegetables, which are important foods for your Energy Plan.

The Energy Plan works by addressing an individual's needs in proportion with their size and weight. Nutrient amounts are considered in grams per kilogram of body weight.

Let's take as examples a female runner who weighs 60 kg and a male pole-vaulter who weighs 90 kg. As set out below, for a rest day or targeting body fat reduction the recommendation is from 2 g of carbohydrate per kg of body weight. For the 60 kg runner this equals 120 g of carbohydrate per day, but for the larger runner it is 180 g of carbohydrate each day. And don't worry – while these figures might appear a bit prescriptive, they are just guideline amounts to give a sense of how intakes change depending on different athletic demands. In

Part II I will show you how to understand your own portion size without needing to do anything complicated or overly prescriptive.

Elite Insight: Carbs

Carbohydrate is the nutrient in which we see the biggest variance day to day, depending on work demands. From the diagram overleaf you can see that on rest days, for most athletes, demands are greatly reduced – from 2–3 g **per kg of body weight**. So our 60 kg runner requires above 120 g of carbohydrate over the day to meet her physical demands. As we will see in Part II, this could easily be provided by two carbohydrate-rich meals using basmati rice, wholegrain pasta or porridge oats.

An elite team-sports athlete – here, a footballer – will for training days increase their intake to 4–6 g per kg body weight. This increases to 6–8 g per kg body weight on match days.

And for endurance performers like cyclists 'fuelling up' in the 24 hours before a race, carbohydrate intakes can increase to above 8 g per kg body weight a day. For Tour de France cyclists on their hardest days in the mountains, intake can exceed a massive 12 g per kg body weight.

For most athletes, it is a big step up from eating to fuel their training to then fuelling up for a competition or event. Consuming the extra fuel necessary takes practice and requires different 'techniques' – there is also only so much food a person can eat, after all. Often elite athletes meet their fuelling demands by using carbohydrate drinks and other sports foods such as gels. For the average person, however, adapting nutrition on different days depending on their goals, the aim is to use real food. The Energy Plan is a food-first approach to nutrition, using meals and snacks before looking to supplements, and I'll show you how to achieve this with the simple planners in Part II.

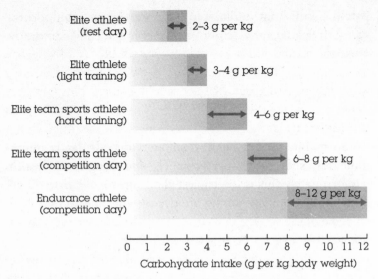

Elite athlete (rest day) 2–3 g per kg

Elite athlete (light training) 3–4 g per kg

Elite team sports athlete (hard training) 4–6 g per kg

Elite team sports athlete (competition day) 6–8 g per kg

Endurance athlete (competition day) 8–12 g per kg

Carbohydrate intake (g per kg body weight)

Athlete carbohydrate intake on different days

Fat: Storage and Protection

Fat provides the body's second main fuel source. Unlike with carbohydrates, our bodies have a large reserve of fat that could, in theory, keep us going for days. Body fat (adipose tissue) is stored all over the body – under the skin, where you can pinch it (**subcutaneous fat**), in the abdominal cavity (**visceral fat**) and within the muscles (**intramuscular triglycerides**).

Visceral fat is harmful, building up as it does around our organs and being linked to an increased risk of chronic health issues such as type 2 diabetes, high blood pressure, heart disease, stroke and even dementia. It's also known as 'active fat' because it can interfere with our hormones.

Our body, of course, needs fat to function properly. Fat plays a key role in helping the body absorb the fat-soluble vitamins A, D, E and K in the intestines, and is an important component of each cell, building cell membranes (the outer layer) and also the sheath around nerve cells. Body fat consists of **essential fat** and **storage fat.** Storage fat is an energy reserve, while essential fat is required for normal functioning of the body. Essential fat is stored in nerve tissues, bone marrow and organs and has a key role to play in blood clotting and inflammation. It makes up approximately 3 per cent of body mass in men and around 12 per cent in women. This large difference is due to the different physiology necessary in women for childbearing.

So maintaining a healthy amount of body fat is important for short-term performance and long-term health. Unfortunately this is the biggest challenge for the over 60 per cent of the population who are overweight, and we will look at some of the strategies you can use to lose fat as part of your Energy Plan, such as training low, in Part II.

Fat is the body's most concentrated store of energy. A gram of fat releases 9 kcal (calories) of energy (compared to just 4 kcal per gram of carbohydrate), and fat is our muscles' preferred source of fuel for lower-intensity activity, such as walking or jogging.

For the body to use it another currency exchange takes place. Stored fat is released into the bloodstream and taken up by those little power generators in the muscle cells, mitochondria, which convert it into ATP. The conversion is done through a process called, scientifically speaking, oxidation (known more generally as 'burning fat', though it's more like melting it down and minting a new currency).

For many of us, tapping into these fat stores and reducing them will form a key part of our goals with our Energy Plan.

Good or Bad Fat?

Fat, like its fellow macronutrient carbohydrate, has been the subject of an avalanche of often extreme and conflicting advice. I'm regularly asked whether all fat is now OK, or if it's wrong to be eating low-fat food. And the sensible answer is that we don't need to deal in such absolutes. Fat plays a key role in helping the body absorb certain vitamins and minerals and is a vital component of each cell, providing invaluable aid in blood clotting and inflammation, which are part of our immune system's response to outside threats. So, just like carbohydrate's role in fuelling our brain and muscles, it's vital to us.

WHAT IS INFLAMMATION?

There are two types of inflammation: acute and chronic. Acute inflammation (its name is slightly misleading as it can last for days) is an important defence mechanism, where white blood cells surround and protect against harmful stimuli and start the healing process, and is vital for injuries, wounds and infections.

Chronic inflammation is when the body can't break down the particular irritant causing inflammation or when someone has an autoimmune disorder. Some fats have an anti-inflammatory effect, while some are pro-inflammatory (see below).

Not all fats are created equally. Some are better for us than others, and it is useful to be able to separate the 'good' fats from the 'bad', to use some of that absolutist language the diet industry is so keen for us to adopt.

Think of your body as a car, and the various types of fat as 'traffic lights' that you might either stop eating, reduce how much you eat or carry on with, even increase your consumption of:

Red: Avoid

Trans fats: traditionally found in margarines (though not so much these days), and in pastries, cookies, pies and French fries. They have been shown to increase harmful LDL cholesterol, reduce beneficial HDL cholesterol and increase inflammation, which is linked to heart disease, stroke, diabetes and other chronic conditions.

Amber: Reduce

Saturated fat: solid at room temperature and common in red meat, whole-milk dairy, butter, ghee, lard, cheese, coconut oil and many commercially prepared baked goods. Limiting saturated fat to under 10 per cent of daily calorie intake is widely advised.[2] Researchers at Harvard University recently found that, although there was not enough evidence to state that saturated fat increases the risk of heart disease, **replacing saturated fat with polyunsaturated** may reduce the risk of heart disease.[3]

Green: Increase

Unsaturated fats: liquid at room temperature and found primarily in oils from plants and fish. There are two types of unsaturated fat: **monounsaturated** and **polyunsaturated**. It's important to understand the foods that contain these 'green light' fats, as food labels mainly focus on total fat or saturated fats. (See the **performance food planners** in Part II.)

A diet rich in monounsaturated fats such as olive, rapeseed and groundnut oil, avocados, nuts and seeds (components of the widely talked about Mediterranean diet) has been associated with a lower incidence of heart disease.[4]

Polyunsaturated fats have a slightly different structure and this group contains some 'essential fatty acids', needed for the normal functioning of our bodies, but they can't be produced by the body (much like 'essential amino acids', which we will come to shortly when we talk about proteins). There are two main types of polyunsaturated fat – omega-3 fatty acids and omega-6 fatty acids.

Omega-3 fatty acids are found in oily fish like salmon and mackerel, flaxseeds, walnuts and rapeseed/canola oil. As well as having positive effects on heart health, there is also emerging evidence on them supporting cognitive function and on reducing inflammation[6] and muscle damage[7] after hard training.

Omega-6 fatty acids – sources include vegetable oils such as sunflower, corn oil and cottonseed oil – are more commonly consumed. It's thought that the higher ratio of omega-6 to omega-3 in many people's nutrition habits may increase inflammation and affect physical and cognitive function. This means making a conscious effort to increase omega-3 intake, which we will return to in the Ageing chapter, page 241.

Research is ongoing in these areas, and as such it's not possible to be as certain with advice as it is with carbohydrate and protein, but it's safe to say that it makes sense to switch the fats in your food to green traffic light sources: mono- and polyunsaturated 'healthy' fats.

IS COCONUT OIL REALLY GOOD FOR YOU?

Although coconut oil has reached peak fashion over the past few years, the evidence doesn't back up the hype. It's high in saturated fat (around 90 per cent), around six times higher than olive oil. So when cooking reach for olive oil instead - and for extra credit go extra virgin for its higher anti-oxidant content. Despite the myths, extra virgin olive oil is suitable for cooking - it's stable when heated,[8] which means the antioxidants don't break down and lose their benefits.

Protein: The Maintenance Job

Protein is a macronutrient that is given a lot of attention in all kinds of dietary advice – and it's little wonder when you consider its role in the body. It's vital to the crucial and ongoing building, strengthening and repair of muscle, bones, skin, hair and major organs, as well as being an essential component of enzymes, hormones and antibodies. It even has a secondary role as a fuel to produce energy (though its role here is far less prominent than that of fat or carbohydrate).

But what are proteins? To put it simply, they are chains made up of building blocks called amino acids. There are 20 amino-acid building blocks, each of which has a different function, and nine of them are known as 'essential' amino acids because they cannot be made by the body – they have to be taken in through food. There are two types of food containing protein:

1. Complete proteins, containing all nine essential amino acids. Examples include meat, fish, eggs, milk, cheese, yoghurt, quinoa, buckwheat, soybeans.
2. Incomplete proteins are plant-based sources that are missing some amino acids and need to be combined with other incomplete proteins. Examples include beans, lentils, nuts and seeds. For example, kidney beans can be combined with wholegrain rice. However, this doesn't have to be done at every meal; the body can use amino acids from recent meals to form complete proteins. Variety is the key here. There are many options for combining incomplete proteins in the food lists in the appendix.

Think of amino acids as building a protein brick wall. You need all of the different types of brick for the strongest, most robust structure. Will it stand up without all of them? Yes, but over time the wear and tear will be more than if you included all the different bricks.

Although the muscles in your body might seem pretty stable, they are in fact in a state of flux. They are constantly breaking down and having new muscle tissue built in a process called muscle protein synthesis (MPS). If you've ever broken your leg you'll know exactly what I mean when I say that, when the cast comes off, your leg resembles a chicken leg. This is because the muscle has reduced due to not being stressed (used) – we call this muscle atrophy. And the fuel to reduce this muscle breakdown while you're immobilised is protein.

So how much protein do we need? The current guidelines for a sedentary person are 0.75 g per kilogram of body weight per day – although hopefully if you're reading this you're either an active person or at least contemplating change, in which case your requirements will be higher. Recent research in this area has increased the awareness of intakes according to programme goals and training demands.

The intakes in the table below range from 1.2 to 2.0 g per kilogram body weight per day for the exercising person.[9] Higher intakes may be required during periods of hard training or energy restriction (reducing body fat[10]).

Sedentary	0.75 g per kg body weight
Endurance/lighter training	1.2-1.6 g per kg body weight
Resistance/ heavier training	1.6-2.0 g per kg body weight[11]

American fitness pioneer Jack LaLanne once said: 'Exercise is king. Nutrition is queen. Put them together and you've got a kingdom.' And protein and exercise enjoy a particularly harmonious marriage. Whether you're an elite athlete, office worker or a retired grandparent, resistance training is the most effective form of exercise to increase strength and muscle mass. It increases the stress placed on the muscles, which in turn requires more amino acids to be taken into them to increase muscle protein synthesis and build new muscle tissue.

In fact, protein is increasingly important for the grandparent, because as we age our muscles become less resistant to the stimulus from training and protein consumption, causing a decline in function and muscle mass called sarcopenia. This is something we'll look at later in the book, in the Ageing chapter.

Dose Up

With your Energy Plan, the timing of your protein intake is important, not just the amount per day you consume. It was once assumed that taking on protein immediately after any kind of training was a priority, but it's now accepted that the protein you eat the day after exercise – for example, for at least 24 hours after a resistance-training exercise – can help enhance your results. And this protein doesn't need to come in single, mealtime doses either. A recent study[12] demonstrated that muscles use protein more effectively if it's consumed in small doses throughout the day – i.e. in the form of snacks – rather than a big single intake, as shown below:

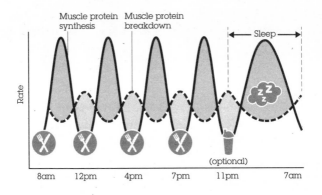

How muscles use protein consumed in small, frequent doses

It's proof that snacks aren't inherently bad and that it's more a question of context (and the type of snacks we're talking about). Building protein-based snacks into your diet can be a positive intervention and, as we'll see in the Recharging chapter, it can help your body's maintenance even while you sleep.

Micronutrients: Your Primary Protection

Micronutrients are nutrition's 'little helpers', involved in energy production and maintenance to keep your body running smoothly. You might know them as vitamins, minerals and trace elements. Listing them all here would take too long, but it's safe to say that they are vital to the functioning of your immune system, which is why we broadly label them as **protection** in the fuel composition for our engine.

Micronutrients are the most misunderstood group when it comes to nutrition, and through the course of this book I intend to shed some light on their importance and give you the tools to incorporate them in your diet to support your body in relation to the demands you place upon it.

For now, though, I'll say that deficiencies can develop if you eat a diet that's too narrow. Even for those who like to cook, when time is short it's all too easy to fall into a recipe rut, churning out the same thing over and over again because it's quick and easy. But eating different grains, pulses, fruit and vegetables (as well as meats, if you're a meat eater) in each meal – eating the proverbial rainbow, if you like – is a good starting position. In the Ageing chapter we will get more specific on how to build your plate with micronutrients.

Active individuals are at risk of having an inadequate intake of some of the most important micronutrients. Here's some more on their properties and the deficiencies associated with not eating enough, plus how to get micronutrients into your diet (RDA = Recommended Daily Allowance):

Micro-nutrient	Function	Sources	RDA
Iron	An important component of haemoglobin (which carries oxygen in the blood). Increased risk of deficiency for distance runners (due to loss from feet hitting ground), women (menstrual losses and pregnancy) and non-meat eaters. Deficiency can lead to fatigue and reduced work capacity.	Haem iron (from animal sources) and non-haem iron (from plant sources). Plant sources such as vegetables, pulses, beans, nuts and fruits are less well-absorbed. However, combining with foods containing vitamin C can enhance absorption.	14.8 mg women, 8.7 mg men. Athletes should aim for higher.
Calcium	Maintenance of bone tissue, muscle contraction, nerve conduction, blood clotting. Increased risk of deficiency if reducing energy intake and/or avoiding dairy. Risk of reduced bone mineral density (BMD) and injury increased further for women with menstrual dysfunction (see RED-S).	Milk, cheese, yoghurt, fortified bread, soya, rice milk, canned sardines.	700 mg Higher intakes required for energy deficiency or menstrual dysfunction (see 'RED-S', page 61).

Micro-nutrient	Function	Sources	RDA
Vitamin D	Maintaining bone and teeth health, muscle function, immune function. Increased risk of deficiency for those with darker skin or limited outdoor access.	Found in a small number of foods, in small amounts: oily fish, egg yolk, liver, fortified breakfast cereals. Supplementation recommended by Public Health England during the winter months (10 mcg, see supplement section). Sunlight exposure recommended to hands, forearms and lower legs for short periods between 11am and 3pm from April to September.	10 mcg
Anti-oxidants	Protecting against cell damage from oxidative stress from exercise and other stressors. There are many classes of antioxidants including vitamins, minerals, carotenoids, flavonoids and phytochemicals.	Fruits and vegetables, which contain different antioxidant compounds, depending on their colour. See the Ageing chapter for full information.	N/A

Water

Water is an often forgotten nutrient, but given that between 50 and 70 per cent of our body is made up of it, it's clearly an important one. Muscle mass contains around 73 per cent water, while fat (adipose tissue) is around 10 per cent water. It's a critical component in cells and helps digest food, transport other nutrients around the body and regulate our body temperature, much like a radiator does in a car engine.

The human body is regulated to maintain a temperature of 37 degrees Celsius, which is optimal for the body to function. When it gets cold the blood vessels constrict the flow of blood to the skin, and when the body's core temperature rises, blood vessels dilate to direct the blood to the skin's surface and remove heat through sweating. The human body has somewhere between two and four million sweat glands, which are triggered by heat and exercise as well as by emotions and hormones – think of those clammy palms before you give a presentation at work.

Sweating is clearly essential, but it can also become a problem, particularly when playing sport or exercising. When we lose more water than we take in, we become dehydrated, and the effects can range from the mild to the much more serious, including in extreme cases, death. Sweat rates during exercise vary greatly, between 0.3 and 2.4 litres per hour.[13] With 1 kg of body weight loss equalling around a litre of sweat loss, fluid losses of greater than 2 per cent of body weight (1.4 kg for a 70 kg person) have shown reductions in both physical and cognitive (decision-making, concentration) function. Evidence has shown that losses of 2 per cent can reduce physical performance by around 5 per cent in cooler conditions, but in hot conditions a loss of 2.5 per cent can reduce physical performance by up to 50 per cent.[14]

So how much water should we drink and how should we obtain it? A common misconception is that all of our hydration needs should come from drinking fluids, but in fact around 20 per cent of our daily needs are met by food. Apples, broccoli, watermelon, lettuce and yoghurt are some of the foods with the highest water content.

Around two litres of water per day for men and 1.6 for women appear to be the accepted guidelines, but don't take into consideration exercise and individuals' sweat rates which can greatly change the amount of water we need. In the Monitoring Your Progress section in Part II I will show you how to individualise your hydration strategy.

Hydration is, of course, very important for the athletes I've worked with. When I travelled with the England football squad to the 2014 World Cup in Brazil, one of the greatest challenges the team faced in their opening game against Italy – played in Manaus, in the heart of the Amazon – was the climate. They would be facing temperatures in the high 30s as well as extreme humidity: a more concise way of putting it was that they would be playing in a furnace.

To address this I flew out a team of experts from Loughborough University to help us monitor every player closely to determine individual sweat loss – we all sweat at different rates, which I'll return to later in the book – and tailor their fluid intake accordingly. But there are much cheaper and just as effective tools to monitor your own dehydration. In Chapter 8, I'll show you how to get a gauge on your own sweat rate so you can rehydrate effectively.

At Arsenal Football Club we had a designer create a colour chart to display in the toilets so the players could see where their urine ranked on the sliding scale of well hydrated to dehydrated, and monitor it themselves. It's safe to say it wasn't the designer's favourite job that season, but it was an effective tool to help the players stay on top of their own maintenance – a little nudge to get them thinking about it. The Arsenal chart scale ranged from pale yellow (most hydrated) to brown (least hydrated). On this scale you should aim for your urine to be no darker than a pale orange or yellow. It should also be clear, not cloudy, and plentiful. If it looks any darker than golden orange you should look to be taking on more fluids, and if the colour is a coppery or darkish brown, you should do so as soon as possible.

What About Calories?

It's about time we talked about calories. Calories, often called kilocalories (kcal) on packaging, are the unit we use to measure the energy that we consume in our food and drink. A single unit (one kilocalorie) refers to the amount of energy required to raise the temperature of a kilo of water by 1 degree Celsius.

The accepted 'average' calorie intake for a man to maintain a healthy body weight is around 2,500 kcal per day, while for a woman the number is more like 2,000 kcal per day. But these needs are affected by your daily activity levels, your age and your metabolism (your engine). A Premier League footballer might consume as many as 3,500 kcal on match day, while a cyclist in the Tour de France might wolf down 8,000 kcal for a day involving a lot of hill climbing.

The first law of thermodynamics states that energy can be transformed from one form to another, but cannot be created or destroyed. So the body is constantly transforming energy from the food we eat to produce heat. Our cells need energy to function and our muscles need it for movement, and each part of our fuel provides different energy levels:

- **Carbohydrates and protein each provide 4 kcal per gram**
- **Fat gives us a whopping 9 kcal per gram**
- **Alcohol, in case you're wondering, isn't far behind at 7 kcal per gram**

With the Energy Plan, our bodies are an engine and our daily food consumption is our fuel for the day. Counting calories alone is overly simplistic, and it doesn't factor in the composition of our fuel. It's important to remember the following message:

Focus on the fuels (the nutrients) and the calories will look after themselves.

We should view calories as our **fuel budget**, the amount we need to expend to meet the demands for the day so that by the end of the day we have neither an energy surplus nor a deficit (unless it's part of a plan).

Our fuel budget will change depending on how active we are on any given day, but within that our focus should be on the nutrients. If you're simply counting calories you can fall into the trap of simply eating *less* or *more*, instead of tailoring your eating to meet your demands. So, if you're eating too much carbohydrate in the form of simple sugars, reducing your portion sizes will reduce this, sure, and now you'll be eating less simple sugar. But what if your body is crying out for more protein? Increasing your protein intake as well as reducing your simple sugars – essentially swapping one for the other – would better fuel your body and meet your goals, and would be more effective than just reducing calories.

The Energy Equation

So now that we have established the composition of our fuel, as well as understanding the engine it powers, it's time to look at the final part of the energy equation. Your fuel is your **energy in**, but it's only once we've addressed your **energy out** that we can hope to find the balance that is at the heart of the Energy Plan.

In the next chapter we'll see how your engine uses its fuel. As ever, if you already have a good grasp of your exercise plan then it may prove a little simplistic for you, in which case please feel free to move on to Part II, where we will deliver your Energy Plan.

For those of you reading on, however, let's look at what happens when we rev our engines.

CHAPTER 3

The Accelerator

There's real poetry in the way the very best performers move. Think of the seemingly effortless grace with which US Olympic gymnast Simone Biles defies the laws of gravity before landing back on earth with a smile, or French footballer Kylian Mbappé gliding past defenders at warp speed at the 2018 World Cup.

Athletes like these have the power to make the hairs on the back of your neck stand up, and it's easy to admire and talk about their ability without taking into account all the planning, training and 'grunt work' that goes into it each and every day, far from the TV cameras and adoring fans. Behind every world-class performance is a schedule that has been put together with pinpoint accuracy to achieve this goal.

So, what can we learn from the elite to use in our Energy Plan and enhance our own training and nutrition? There's only one place to start, with a simple-seeming question:

Why Am I Doing This?

The philosophy of clearly thinking through and detailing why we're doing something rather than jumping straight into it works very well with nutrition and training. Every athlete I work with has their 'why?'. They have specific goals to meet and their training and nutrition are tailored to meet these goals.

In a sport like football, it's easy to have the perception that the team just trains together each day, working on technical elements of the game such as attacking and defending, as well as fitness work. But the reality is that all of the players are assessed as individual athletes to determine what can be done to improve performance on the pitch and ensure that this can be repeated as many times as there are games over the season. So across a squad, each player will have different goals – each athlete starts with their own. They are able to answer:

- What is your goal?
- Why are you using a particular type of training?
- Does each of your training sessions contribute to the overall goal?

As soon as there is a goal you can write down, all of your training and nutrition can be aligned to meeting this. We will look at how you can go about establishing your own goals as part of your Energy Plan at the beginning of Part II. For now, whether you're lacing up your running trainers for the first time in forever or you're well into an established exercise regime, I want you to start thinking about your own 'Why am I doing this?'.

Training Principles

The physical activity that makes up part of your Energy Plan – the part of your 'energy out' that we can exercise most control over, as we discussed at the end of Chapter 1 – will inevitably involve some of the following types of training to allow you to make the adaptations necessary to reach your goals. It's likely that your current regime already involves at least one of these; for many people it's aerobic exercise.

Endurance (aerobic) Exercise

Often described as cardio (short for cardiorespiratory), aerobic exercise causes an increase in heart rate, which pumps oxygenated blood to the muscles and uses fat as its source of energy. It typically describes lower-intensity exercise such as walking, jogging, cycling, swimming and the various cardio machines in the gym.

Aerobic training improves the delivery of oxygenated blood to the working muscles to produce ATP (as discussed in Chapter 1). The heart becomes more efficient, increasing its stroke volume and reducing heart rate, and within the muscles it increases the density of capillaries to deliver oxygen. It also increases enzyme activity and the number of mitochondria – those power generators that melt fat down and change it to your muscles' energy currency, ATP – meaning more fuel, specifically fat, can be burned to produce energy at lower exercise intensity, improving exercise capacity.

As soon as the intensity of exercise increases it crosses the **anaerobic** or lactate threshold. At this point oxygen can't be delivered to the muscles quickly enough to meet energy needs, and becomes 'anaerobic'. At these higher intensities the body starts to use more carbohydrate as a fuel, but by-products such as lactate and hydrogen make the muscle cells more acidic. For many elite and sub-elite runners, a focus is to raise their lactate threshold and allow them to use more fat 'aerobically', which both preserves valuable carbohydrate fuel (for a sprint finish, for example) and reduces acidity within the muscle, which can interfere with muscle contraction and cause fatigue.

Strength (resistance) Exercise

Often referred to as resistance exercise, strength training is when stress is placed on muscles against a given resistance, and the contraction causes micro-tears in fibres (otherwise known as muscle breakdown), which subsequently grow and repair, resulting in increases in strength and muscle mass. An increase in the size of muscle fibres (which has the grand-sounding name of hypertrophy) is due to an increase in size and number of fibres.

Resistance training is a crucial component across almost all elite sports to increase strength and power, allowing the muscles to produce greater force – whether this is a sprinter bursting out of the blocks or a golfer on the first tee with a driver in hand.

Working on your muscles not only makes you stronger, but also stimulates bone growth and repair – muscles feed off blood glucose and muscle glycogen stores, lowering circulating levels, and there is an increase in metabolic rate thanks to increased muscle mass. Increased muscle also makes the body more efficient at burning fat, and improves balance and posture.

Intermittent (concurrent) Exercise

Elite team players in sports like football train to improve both strength and endurance adaptations, to support the repeated sprints needed over the 90 minutes of a match. We call this concurrent training and it means, in general, that on-pitch training sessions and gym work are separated to minimise the 'interference effect', which reduces the effectiveness of the training for both objectives. Intermittent training also includes the popular HIIT, which combines high-intensity exercises with short recovery periods within the same session.

Of course, we shouldn't forget the other types of training such as flexibility (mobility), which is an important part of warming up and down after training, and neuromotor exercises such as bounding lunges or hops, which form a part of many training routines. Yoga and Pilates are particularly effective forms of flexibility exercise.

As we will see during this chapter, what you eat before or after these different training sessions will either enhance or reduce how your body adapts (your gains) from the training session.

The Low-hanging Fruit: Incidental Activity

If you've ever seen a sprinter train or perform at an Olympic Games, you'll know it's quite a sight. They explode out of the starting blocks with such power, and move off at a rate that you might have thought only belonged in wildlife documentaries filmed in the Serengeti.

Because sprinters work at such a high level of intensity, they need sufficient rest between efforts to allow their body to recover. Get the balance right and you have an athlete primed to perform at their best in competition. Prior to the Olympic Games in Beijing, however, we weren't quite getting it right with one sprinter. No matter how many changes we made to his nutrition, we couldn't quite get him lean enough. So we sent him out to Jamaica to do some training with Usain Bolt's coach Glen Mills, and he came back a different athlete, having shed some four or five kilos of body fat. So what had he been doing differently? He'd increased the volume of his activity.

His training sessions had expanded from just doing the explosive work with rests in between to doing more base conditioning (running longer distances), so his overall daily energy output increased. In other words, *he stopped being a sedentary athlete.*

When I work with any client it's important that I not only see an account of the food they eat (energy in), but also learn about their exercise or training regime (energy out). They usually fill in a form, which my team then analyse, and without fail there is always one thing missing: incidental activity. Our brains are so hard-wired to see energy output as 'formal' exercise – yesterday's HIIT session, our morning swim – that we miss the low-hanging incidental fruit.

What I will usually do is ask to see the client's phone – I want to know if they have a health app that tracks their step count, and whether they've been taking the stairs or the lift.

I know, the data on your mobile phone or tracker won't be 100 per cent accurate, but it doesn't need to be. Increasing your daily step count from 5,000 to 10,000 steps can be a significant step towards creating a sustainable energy deficit to lose body fat; and that's before you've even addressed your nutrition. It doesn't just have to be death by treadmill.

Increasing incidental activity can involve something as simple as walking home from work instead of catching the bus or taking the stairs instead of the escalator. It all adds up. Obviously it's easier to decide to walk home on a lovely warm spring day, but the real test comes in the middle of winter, when it's cold and the rain is lashing down. Think about the benefits your incidental activity brings, wrap up warm or take your umbrella and walk home anyway.

In order to achieve a balance in your physical activity, it's useful to treat training and incidental activity as two separate entities, and break them down into 24-hour periods. So, when training is **high**, incidental activity levels outside of training should be **moderate**; when training is **low**, activity levels can be **increased** to maintain volume. In a way we're not only looking to balance our **fuel in** with our **energy out**, but also to find the right balance to maintain the necessary levels of energy out so you can work towards your goal. This is the same principle used for planning athletes' training – manipulating the intensity and the volume.

Filling the Tank

There's no point going into a heavy training session with an empty tank, just as there's no point in looking good if you can't deliver a performance. With our athletes, the main focus of their nutrition programmes is to prioritise what they eat before training or performing (**fuelling**) for energy supply during said training and performance, and then what they eat after, for **recovery**. With your Energy Plan, the same principles apply, with the **type** and **timing** of your food, which we'll return to at the end of the chapter, being of paramount importance.

The first question to consider is where do you sit on the continuum below: are you **adapting** or **performing**?

This is the question the coaching staff and I consider with each sports team I work with because it dictates what the athletes eat before the training session. With a football club during pre-season training, carbohydrates are often restricted before some sessions to improve fat metabolism. This is with the understanding that it will feel harder for the players, so the session intensity will be lower.

What to eat before your workout depends on whether your objective is to adapt (improve endurance adaptations, such as fat burning) or to perform (ensure sufficient energy supply) to maintain intensity. Your training week will often include the need for both, depending on your goals. All of the examples in this section give an insight into how athletes fuel pre- and post-training. How to personalise your fuelling plan to meet your goals is covered in Part II.

PRE-TRAINING 1: Performance

The pre-training meal should be scheduled between two and four hours before training to allow time to digest your food. As the goal is **performance** to maximise your output in the training session, you will need to be fuelled (training before breakfast as part of adaptation is below). Examples of meals are included below. Snacks may also be included, less than two hours before training, depending on the time of training in relation to your schedule (if training is late morning or late afternoon, for example – a long time after breakfast or lunch).

For competition, for professional footballers, it is recommended that they have the pre-match meal three to four hours before kick-off, usually at the hotel before travelling to the ground. Snacks containing high-GI carbohydrates are then consumed during the warm-up to top up blood glucose. A meal containing 1–3 g of carbohydrate per kg of body weight is the standard practice (and see the competition section in Part II). So for a hypothetical 60 kg midfielder, a meal containing around 1 g of carbohydrate per kg of body weight (60 g) could be a bowl of pasta.

EXAMPLES OF FUELLING MEALS

- Porridge oats with milk of your choice
- Quinoa porridge with chopped banana and milk of your choice
- Tomato and basil omelette with rye bread
- Mexican beans and avocado on rye bread
- Homemade muesli with mixed nuts and seeds
- Chicken or tofu stir-fry with noodles
- Baked sweet potato with tuna
- Cajun chicken with avocado and quinoa salad
- Vegetarian chili with rice
- Wraps: (salmon and avocado, chicken, falafel)
- Buckwheat salad with prawn, chicken or tofu skewers
- Spaghetti with meatballs or sardines

FLUID INTAKE GUIDELINES

Guideline recommendations are to consume **5–7ml of fluid per kg of body weight in the two to four hours before training** (with the pre-training meal is an ideal opportunity). This will ensure any excess fluid is passed before the start of the session. So for a 70 kg person, we're looking at between 350 and 490 ml. As with all the measures in this book, you don't have to measure them out to the nearest gram or millilitre: with time, you'll get a feel for the right amount, and I'll show you how to get better at estimating these amounts in From Plan to Plate, Chapter 10.

PRE-TRAINING 2: Adaptation

For some training sessions, the aim is to train the body to adapt and become more efficient, rather than to perform at full capacity. A primary tool to achieve this is called 'training low'. As we've discussed, there is a vast reservoir of fat in your body, in tissues, cells and your blood... and for many of us, that reservoir might be a little *too* large. Our engine has a wonderful ability to use two fuels, switching between carbohydrate and fat depending on the intensity of activity. This ability is called '**metabolic flexibility**' and it means that your metabolism uses different fuels depending on the demands placed on the body. And where there's flexibility there is, of course, the opportunity to manipulate it to our own ends. In my work with elite athletes I encourage and teach them to match their eating to these demands, and here we'll look at how you, whether you're in training or you just want more from your exercise classes, can use this flexibility too.

I'm often asked by clients how to prime their body to burn more fat. And the answer is that you restrict carbohydrate. There is a *but* coming, however...

Marathon runners and triathletes have been using this approach for years, because they want to make their body use fat as a fuel and preserve the limited stores of carbohydrates for when it's 'eyeballs out' time and they need that quicker currency of energy for a sprint finish. If they get it right, their engine becomes very efficient at using both fuels; but get this wrong and they'll hit the wall as the finish line approaches, all jelly legs and expended effort. (In Part II of the book, I'll show you how to get it right.) Recently, low-carb high-fat (LCHF) or keto diets have received a lot of attention, and though this may be useful in some weight management scenarios, the ability to use both fat and carbohydrate as fuels during different sessions is important to maintain exercise performance.[1]

When muscles' glycogen stores are reduced and the supply of carbohydrates restricted through training low, they become more efficient at using fat stores as fuel. There are various ways to train low:

- Exercising twice a day
- Prolonged training (over 90 minutes)
- The dreaded low-carb diet
- Exercising before breakfast, the most popular way. However, recent research has shown that training after a low-carb, protein-based meal will have the same effect.

TRAINING-LOW MEALS

- Greek yoghurt and handful of nuts and seeds
- Omelette or eggs (poached or scrambled)
- Smoked salmon and avocado
- Mexican spiced beans and avocado
- Whey protein shake

It should also be noted that training low isn't a magic bullet to burning fat. It increases the strain on muscles and bones and your immune system when exercising, and can reduce the quality of harder sessions. It needs some careful planning and should be first trialled on shorter, lower-intensity training sessions. Also, as we have discussed, this is focusing on the fuel used during training, and to reduce body fat you will need to be in an energy deficit over the course of a day.

Mobile Fuelling

Sometimes it's necessary to take on fuel *during* activity too, depending on your goal so here are some guidelines:

Exercise Length	Carbohydrate Amount
< 45 minutes	Nothing
45–75 minutes (high-intensity)	Small amounts
1–2.5 hours (endurance)	30–60 g/hour
2.5–3 hours (ultra-endurance)	>90 g/hour

Adapted from Thomas et al., 2016

30 g of carbohydrates are provided by the following:

- 500 ml sports drink (isotonic)
- 1 carbohydrate gel
- 1 large banana (if you're a tennis fan, you've probably seen a lot of players eating these during matches)
- 1 sports nutrition bar
- 1 cereal or breakfast bar

These are all high-GI (easily absorbed) sources, for when high-GI is a good thing for a quicker spike in energy.

For most people, taking on extra fuel during training isn't necessary. For example, in football taking on carbohydrate during the match is advised, as the goal is to be fuelled and maintain energy supply and performance levels. But carbohydrate isn't used during training sessions; instead players are fuelled before the session. For harder training sessions, you should do likewise, fuelling beforehand in the form of a carbohydrate-based meal.

ENDURANCE NEEDS

If you are training for a marathon, Ironman or other endurance events, some of the sessions will involve training low so your body can better adapt to using fat as a fuel, but leading into a race it is important to practise consuming carbohydrate during longer training sessions (to practise your **competition strategy**). This is because it is possible to '**train the gut**' so that consuming carbohydrate during a longer run increases the absorption in the intestines and then oxidation (use) for energy.

This should start around eight weeks before the event, when you are bringing together your competition strategy. Long weekend training sessions are a staple for many preparing for an endurance event, and this longer run (or bike for cyclists and triathletes) provides an excellent opportunity to practise this strategy.

Starting with smaller amounts - around 30 g per hour - and building tolerance is a good approach. Usually the biggest issue on marathon day is GI discomfort from taking on too many carbohydrate products without previously using them.

Fluid intake during a training session should offset sweat losses and prevent you from getting dehydrated. As individual sweat rates can vary hugely, depending on session intensity, duration, fitness and environmental conditions, you will need to personalise your hydration strategy, which you can do in Chapter 8 on monitoring.

Recovery: Post-training

For athletes the period directly after training or competition is crucial, especially when there is limited time before the next training session or match. And it's the same in your own life. Applying **the 4 Rs of recovery** can mean the difference between skipping into work the next morning or limping in, grimacing every time you walk up the stairs.

Refuel glycogen stores (around 1 g per kg body weight per hour for up to four hours, depending on intensity). For a 60 kg professional midfielder this is two large bananas or a plate of pasta each hour for up to four hours; for most of the rest of us, one fuelling snack or meal afterwards – see Chapter 10).

Repair muscle tissue and stimulate muscle protein synthesis (around 0.3 g per kg body weight) – a 20–30 g dose such as a pint of milk (for more options see pages 266–7).

Rehydrate and replace fluid losses from sweat (drink 150 per cent of sweat losses from exercise).

And finally **Rest**.

Refuelling is top of the list for a reason. During the first two hours after exercise, glycogen synthesis (your body's ability to refuel) is at its most rapid – around 150 per cent of normal rate – so it's no wonder that this has been termed the 'window of opportunity'.[2]

Elite athletes use sports food such as carbohydrate and protein recovery drinks directly after performing, followed by a meal in the changing rooms an hour later. Higher carbohydrate intake through meals and snacks eaten every three to four hours is then maintained for the rest of the day, to fully refuel [see competition day in Part II].

These regular meals and snacks should include protein, as discussed on page 35, to maximise muscle repair and adaptation (muscle protein synthesis).

Context is important here. The bigger the dose of exercise, the more important the recovery nutrition. For a lighter training session, which hasn't drained your fuel stores (think about the fuel gauge in the previous chapter), you don't need to add extra recovery drinks or snacks – a meal containing carbohydrate and protein is sufficient. It is important that training is organised so that the next meal can also be used for recovery (so for example, dinner would be the recovery meal for a late-afternoon training session) – especially when reducing body fat and keeping to a daily energy budget (we'll explore this more fully in Chapter 10).

EXAMPLES OF RECOVERY MEALS

- Couscous salad with meatballs
- Mediterranean fish parcels with ginger, spring onion and sweet potato
- Hot and sour fish soup
- Mexican stew with quinoa and beans
- Black bean, tofu and avocado rice bowl
- Fettucine with beans or chicken
- Spinach, sweet potato and lentil dhal

Do I Need a Rest Day?

Adequate rest is critical to continued progress, and a period of at least 48 hours is recommended between resistance-training sessions to stimulate the adaptations in muscle cells for hypertrophy and gains in strength.[3] We know that the muscle is more sensitive to protein intake in the 24 hours after training, so what is eaten over this period has the ability to enhance training gains – it's that powerful.

Rest days are also important for both a psychological and a physiological break. And the dangers can creep up on you if you get the balance between stress and recovery wrong. This is something we know all about in elite sport, where athletes are often treading that fine line between training stress and recovery.

Feeling fatigued in the 24–48 hours after training is normal (sore muscles, reduced energy levels), but when this continues for longer it starts to be more of an issue.

First comes 'overreaching'. This is when it can take a couple of days to fully recover from harder training, to adapt and come back stronger. It's more severe when this crosses into a syndrome called overtraining, which can last for months. The changes to the body's physiology may include increased resting heart rate, increased muscle soreness, hormonal disruption (e.g. low testosterone, high cortisol), weight loss and reduced immunity and appetite.[4]

Bring It Together: The Energy Equation

So we've started to look at how to manage your eating around your training sessions, which with the Energy Plan is part of a broader question examining your overall energy balance over 24 hours. On some days you will have no training, of course, and on others you will have harder or lighter sessions. Finding that balance between your fuel in and your energy out is the key to reaping the benefits of your Energy Plan, enjoying sustained energy to meet the challenges in your life and to look and feel your best. But it's important to get the balance right, and not to let the scales tip too far in one direction or the other.

Positive Energy Balance

A positive energy balance sounds like a nice place to be, but don't let the word 'positive' fool you: depending on your goal, it could be a negative result. This is where you have an excess of energy, after consuming more than your body has used. You've filled the car up and taken her for a ride, but at the end of the day there's still plenty left in the tank.

A limited amount of carbohydrate can be stored as glycogen in the muscles and liver, but any further excess, along with excess fat and protein, all suffer the same fate: they are stored as fat. Surplus carbohydrate and protein can both be converted into fat for storage (the process doesn't work in reverse, nice as that would be).

There is no protein store for the body to dip into; protein is instead contained within your muscles, liver and other tissues. So if you're on a resistance-training programme to increase your muscle mass, a positive energy balance can be a good thing for you as, provided you have sufficient protein intake within that, it will help build new muscle tissue.

And for those for whom a positive energy balance isn't a good thing, it's important to remember that you don't put on 10 kg overnight. It comes from days, weeks, months and years spent in a positive energy balance. It's habitual, something we fall into without any real conscious thought. We have the same breakfast, go to the same sandwich shop at work for lunch, grab those same ingredients from the supermarket on the way home. Our nutrition becomes rigid as we blindly follow one plan. In Part II, we're going to look at how you can break those habits and develop new, sustainable practices with your nutrition.

Negative Energy Balance

Once the body is expending more energy than it's taking in, weight loss should – theoretically – occur. But it won't necessarily be the kind of weight loss you want, as you could be losing anything from body fat to muscle glycogen, water bound to glycogen or muscle mass. Much of the initial weight loss following a low-carb diet is likely to be a reduction of fuel stores – stored water and glycogen in the body – not body fat, as you might be hoping for.

Losing weight in the short term isn't too difficult. There are a host of weight-loss diets on the market that will help you do this, and all of them work on the same principle: putting your body in an energy deficit. However, few of them can realistically promise that they are sustainable – ultimately that would be bad for business, after all! – and it's important to realise that being in an energy deficit for a prolonged period of time is not enjoyable or healthy. If you only put a little fuel in the car and take it out for a long ride, you'll be running on fumes with the fuel gauge screaming WARNING! And these warnings now come with a name...

RED-S

Relative Energy Deficiency in Sport (RED-S) is a relatively recently identified phenomenon. It was previously known as the Female Athlete Triad, and was, as the name suggests, thought to occur only in women when there was insufficient energy available for the body. This was known to lead to irregular menstrual cycles (amenorrhoea), leading in turn to a decrease in oestrogen and other hormones and a low bone mineral density (BMD). Low BMD means an increased risk of injury, often in the form of a stress fracture thanks to the weakened bones, but also has broader consequences for the body and mind.

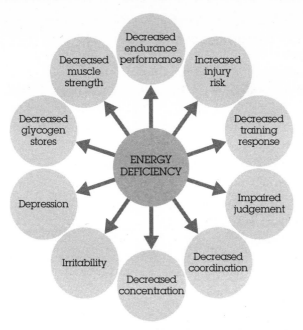

Adapted from Mountjoy et al. 2014[5]

The problem is now known to be no longer confined to women – men are susceptible too – and it isn't just athletes who suffer from it either. If you're running on an energy deficit and you're pushing too hard, you might be in trouble.

Think about this. You're currently in the middle of your exercise programme, pushing hard whilst restricting your energy in – but how do you know when you've gone too far? The diagram overleaf, from a recent paper in the *British Journal of Sports Medicine*, describes the symptoms of being energy-deficient.

If you've ever been on a diet in the past, do any of these sound familiar? In Part II of the book we will explore how you can achieve your goals without suffering these consequences.

The TTA Model

Having looked at some of the principles around when and what to eat when exercising, we now look towards Part II, the more practical section of the book. It's time to introduce the TTA model, which features the core principles of performance nutrition:

- **Type** of fuel
- **Timing** of food in relation to exercise
- The total **Amount** per meal or snack and over 24 hours

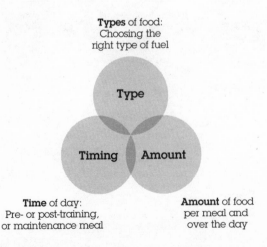

Types of food:
Choosing the
right type of fuel

Type

Timing Amount

Time of day:
Pre- or post-training,
or maintenance meal

Amount of food
per meal and
over the day

These three overlapping factors are the backbone of using nutrition as part of your Energy Plan, and we will now put them into practice.

PART II

Your Energy Plan

CHAPTER 4

Getting Started

What do you have in common with a professional boxer, a Premier League footballer and an Olympic sprinter? Not a lot, you might initially assume – but think again. What do you hope to achieve with your Energy Plan? Your goals might not be a million miles away from those of the performers I work with:

1. The boxer's goal is to slowly reduce body fat in the lead-up to a fight, so they make their weight category (e.g. lightweight or middleweight). They can do this quickly as the weigh-in approaches, through taking on less fluid and food – effectively going on a crash diet – but what impact will this have on their performance?
2. The footballer needs an excellent standard of fitness and energy to perform in a match, often twice a week.
3. The sprinter works on their strength to have the necessary explosive power for their short event.

Their goals may be very specific, but in broad terms they stand for the same as many of our own: we might, like the footballer, want to **improve general fitness**, no matter what our starting point is; or, like the boxer, we might want to **lose body fat**; or we might be more the sprinter type, looking to **improve strength and power**.

And the broader goal for each of them, of course, is to **have the energy to perform at their peak.** For the boxer, there's little point making the weight category if they're going to get knocked out in the first round.

In order to view nutrition through the right lens, whether you're a top performer or simply someone becoming alive to the idea of living a less sedentary lifestyle, it's important to start with your goal. Think how much easier your day at work is with a to-do list or a trip to the supermarket is with a shopping list. The goal could be to reduce body fat, to maintain energy levels during a nightmare week at work or to improve your strength in the gym. Once you have your goal you can begin to tailor your activity and your nutrition towards it, instead of, as it were, blindly wandering the aisles.

When I work with any team or individual, the first thing I do is establish the goal; then I can put in place a process to help them get there. And through establishing your own goal you can make your first steps to cutting out some of the noise surrounding nutrition, and begin your Energy Plan in earnest.

Getting started with an exercise routine can be difficult, as can tailoring your current exercise routine into something with greater purpose. And even once you take the first step, it's easy to fall into the trap of going for that same old run around the block – not too fast or too slow – on autopilot, or wandering around the gym with the anxiety that comes from wondering, *Which machines should I go on today?*

Even if you have your exercise sorted, thank you very much, then there's your nutrition. There are thousands of foods that contain positive health benefits for you. Things like manuka honey, turmeric, cherry juice... or whatever else the self-appointed 'wellness guru' at work is waxing lyrical about. The food industry is seemingly set up to maintain this conveyer belt of elixirs for our body and mind. But *should* you be consuming these? After all, a dollop of coconut oil here, some nuts there and maybe a tasty flavoured kombucha drink to wash it all down can add a substantial number of calories to your intake – and might actually be sabotaging your progress.

These elements might all form part of an approach in which your goal is to 'exercise more and eat better'. But I've got some bad news for you: you haven't established a goal at all.

The language we use is paramount. 'Exercise more and eat better' is too grey, too fluffy – what would meeting this so-called goal even look like? What would it *feel* like? Why would it matter to you?

Why the Energy Plan?

The Energy Plan isn't a diet; it's a journey. My hope is that you will adopt some, many, or all of the approaches within these pages and that you will still be using them not just over the next few weeks, as you might with a diet, but over the next few years and beyond. With my clients and performers, I work to upskill them about their body and to answer all the challenges they will face – fat loss, muscle gain, recovery, immunity, sleep – so that they then have the tools to problem-solve for themselves. You'll be able to do the same with your Energy Plan. I regularly have clients coming to me who say:

- 'I want to feel more energised.'
- 'I need to lose weight.'
- 'I want to run faster.'

Naturally, I want to know *why?* In my experience, just wanting to have more energy or lose weight isn't enough. They're not meaningful enough.

- 'I've just landed my dream job but I need to be able to maintain my energy levels as I'll be working longer hours.'
- 'I want to reduce my body fat, as I'm getting a bit older and have a young family now and I want to be as healthy as I can be.'
- 'I want to run a personal best at the London Marathon.'

All of a sudden these goals have come to life and I can understand the deeper reasons and motivations behind them. The language is so important, and, even though it might take a bit of deeper digging to get there, it's worth doing it so you can set out something serious and with greater intent.

Can you identify your 'why am I doing this?'? It's vital that the hook is meaningful enough to you. Will it get you out of bed on a cold November morning?

Good Intentions

Away from elite sport, many of the people I work with fall into two categories:

1. Those who enjoy training and are looking to improve and reach a goal (such as running their first marathon or improving their cycling performance).
2. Those either exercising or **intending** to get more physically active because they want to be healthy, perhaps reduce body fat, feel more energised and offset long-term health complications.

If you're in the latter camp, if you're at the beginning of wanting to become more physically active and eat better, or you're trapped in a bit of an exercise rut, perhaps it's time to meet one of my clients:

Simon is a music producer. He travels all over the world, meeting and working with some of the brightest talents all the hours he can to maintain his position in the industry. And if there's one thing his job does not offer, it's a balanced lifestyle around exercise, travel, sleep and personal time.

Simon isn't in the best shape of his life, and when he's not spiked by caffeine, he feels tired and sluggish. He likes the idea of doing something about it, and occasionally makes it

*to the gym for some scattergun sessions on the treadmill and
weights, but the language he uses when we meet is telling:*

'I should do more exercise.'

'Managing nutrition on the road is tricky.'

*As the conversation evolves, the roadblocks mount up
in front of him. And, once we finish our coffees and say our
goodbyes, I don't hear from him again. I don't expect to. He
isn't ready to make a change.*

Fast-forward two years, and I receive a text: James, can
we meet this week? Simon.

*When we meet, Simon looks no different – but everything
has changed. He's recently become a father for the first time,
and if he thought he didn't have enough energy before...*

*He's struggling with balancing the new demands at
home with work, and the sleepless nights and pinched
time have got him into some eating habits he feels he
needs to kick. He badly wants to have enough energy to be
an engaged dad and partner. He wants to be around long
enough to see his daughter grow up.*

When Simon came to me a second time, he had **intent**. He was
serious about beginning the Energy Plan, and whether you're
a finely tuned fitness lover, someone who's beginning to think
about dusting off their gym kit or anywhere in-between, as
long as there's a desire to engage with being more physically
active, then the Energy Plan can work for you.

Stepping Up

For those at the beginning of the journey, stepping on to the
first rung of the Energy Plan ladder can be as simple as boosting **physical activity** with incidental activity (see Part I), being
physically active and moving the body more each day. If you're
currently a sedentary person wanting to make a change, this is
where it all starts for you.

As you step further up towards the middle of the ladder, you move on to **exercise**. This involves revving your engine in the form of some aerobic or resistance exercise, and is a good and necessary progression from basic physical activity. But it's also the stage many find themselves plateauing in, doing some exercise but not getting the maximum out of it.

At the top of the ladder is **training**. Targeting goals that are specific and meaningful to you as part of your Energy Plan demands that you step up from your mindset of 'doing some exercise' to one of training to meet your goals. If you are planning to increase strength and muscle mass, then prioritise resistance training sessions in your training week. If you are aiming to reduce body fat, ensure there are sufficient aerobic sessions. There is more intent here. Professional athletes aren't 'doing exercise'; they're training to meet their individual and team goals, and there is no reason why you should not be taking your goals just as seriously. You'd be amazed at just how powerful a slight change in language and mindset can be. Ask yourself:

1. *Why* are you training (or exercising, if you can't quite bring yourself to call it that just yet)? What is your goal?
2. Do each of your training sessions contribute to the overall goal?

As soon as there is a goal you can write down, all of your training and nutrition can be aligned to meeting it. And if you already have a strong goal and exercise regime, now might be the time for you to move on to the next chapter, Performance Plates, which will be the main part of your nutrition strategy as part of the Energy Plan. For those still struggling with exercise, read on:

Roadblocks: But I *HATE* the Gym!

There are still large swathes of the population who don't engage with exercise – and a lot of people who would *like* to, but can't find an approach that works for them, whether that's because of time constraints or just feeling that they don't know where to start. It can be daunting, and a big part of that stems from the fact that a lot of people don't like exercising. They don't like training or the gym. And I would say to them, that's perfectly normal!

It's quite easy to be put off by people in the gym with incredible physiques taking selfies while they look like they're having the time of their lives. For most of us this isn't a reality, and even with the athletes I've worked with it isn't necessarily the case. I've worked with professional footballers and track-and-field stars who absolutely hate the gym. They love the sport they perform in, and being at their absolute physical peak is part of the job, but they just loathe the work in the gym. It's not for everyone – and I'd include myself in that.

I've tried countless types of gym and training programmes, and found I never liked spending hours on them. It's a conflict because while, thanks to my work in sport, I understand better than many people the importance of resistance-training programmes, I just don't enjoy lifting weights in a gym to do it. However, when I tuned in to some of the language used in my work with professional footballers I began to see training work in a new light:

Have You Taken Your Dose This Week?

The strength and conditioning coaches at Arsenal would ask, 'What is the minimal **dose** of resistance training required to keep strength and power levels at their peak?' We call this the **minimal effective dose (MED)**, and it was a game-changer for me. Instead of spending over an hour in the gym, the coaches worked out that gym work could be reduced to 20 minutes after the players had finished training on the practice pitches, working on the core exercises to keep the players strong. The result was more productive sessions and a happier bunch of players – although the key (or catch) here is that if you are in the gym for a shorter period, you need to be prepared to push yourself harder while you are there. For example, taking your muscles to fatigue at the end of a set of repetitions.

Healthcare professionals are starting to prescribe exercise as medication, certainly in the case of those conditions that can be treated through more physical activity, and in this sense using terms associated with medication like 'minimum dose' fits perfectly. The idea of a minimum dose of exercise lowers the barrier to entry. It's the difference between thinking you haven't got an hour free to go to the gym before work and deciding that, actually, you can make time for a 20-minute session. It will help reduce your stress levels and give you more energy (indeed, recent research has highlighted that this can be done in just 13 minutes).[1] If there's a shorter way to achieve what you need, why would you not use it? Just as with your nutrition, there's no point doing a little of every kind of exercise you think is good for you if it's not helping you reach your goal. I know that when I go to the gym now, it's for a 30-minute session and then I'm out the door, and I feel much better about it. If I'm in there anywhere near an hour something's gone wrong somewhere.

Finding your dose can take time. It might involve getting some help from a personal trainer or a friend or colleague. Or it might mean finding something sociable, like a fitness class, running or cycling group (you can get your dose from aerobic exercise as well as from resistance work), that makes you feel like exercising isn't something separate from your life that you have to do reluctantly. In Chapter 8, Moving the Needle: Monitoring Your Progress, we will look at how to measure the progress you are making in your Energy Plan. By experimenting with various kinds of exercise, you'll be able to see which is delivering the best results – without leaving feeling that you *HATE* this! Because you're not going to carry on doing something you loathe that much for ever.

Build Your Base

One piece of advice from my strength-and-conditioning colleagues rings true: build your base. This applies whether we're talking about a runner building aerobic fitness before increasing pace and adding higher-intensity sessions, or going to the gym and learning how to move and use the correct technique before adding resistance or weight. In the rush to reach your goals it can be tempting to jump into the hard stuff, but this can often lead to hitting a brick wall. Remember, the Energy Plan is a long-term strategy, so putting in the foundations at the start will allow you to enjoy its benefits more fully for longer.

Pulling It All Together

With a goal established and our training planned, we can now look at bringing it together with fuelling yourself to meet these demands – your Energy Plan. Over the course of this part of the book we will look at how to build your nutrition plan, starting one plate and then one day at a time and building up to a week. We'll look at the drinks at your disposal, from fuel injections like caffeine to potential gremlins to the system like too much alcohol. We will look at how to manage your environment and stock your cupboards, instilling the winning behaviours that will give you every chance to meet your goals. We will introduce the tools to monitor your progress, the checkpoints to see where any bumps in the road may lie and the impetus to help you adjust to meet these challenges, avoiding giving up at the first sign of difficulty.

And if you've been guilty of consuming too many healthy foods as part of an 'exercise more and eat better' approach, then a useful mantra to remember through this part of the book is: *If it's not helping you reach your goal, what is its role?*

So, let's get started on the fuel of your Energy Plan – building your Performance Plates.

CHAPTER 5

Performance Plates: Different Fuels for Different Days

My job at a football club is to coach the players to be able to modify their nutrition each day according to their demands as they work towards their end goals. The demand for fuel in the form of carbohydrate, for example, is much higher on match day than it is in a training or rest day, and their nutrition is tailored accordingly. And this is the foundation upon which the Energy Plan is built: *The demands you place upon your body each and every day are different, and so you must fuel your body **differently** every day to meet these demands.*

Eating the same things every day out of habit – toast for breakfast, the same sandwich from your favourite sandwich shop, going home late and ravenous and having the same go-to stir-fry – isn't going to meet your Energy Plan's requirements. Following an unchanging diet plan is like putting exactly the same amount of fuel into your car's engine every day no matter whether you're making a short journey or a long-distance odyssey. It just doesn't make sense. So let's take a look at how we address these habits with the elite athletes I work with.

At most Champions League football clubs, a player will report to the training ground before training, either in the morning or the afternoon, depending on when the next match is. When they report in the morning, the TV screens that adorn the walls of the changing rooms and around the training ground display the training plan for that day. And when they go to the restaurant at the ground, waiting for them on the TV screen there is the plan and strategy for today's meal (or meals) and how the players should build their plates to ensure they get the right fuel.

This may sound quite extreme, but it's all part of a strategy to ensure that each meal is linked to the overall goal. The players' food is served buffet-style – they go up and choose the food themselves – as part of our push to educate them on their own eating habits. We want them to take ownership for this because, while they might eat up to 480 meals a year at the training ground, what happens during the other 612 meals they will eat at home, on international duty or in restaurants?

Through this approach they will pick up a process they can use every time they have a meal; this education by habit at the training ground will allow them to use the process anywhere.

As we'll discover during this chapter, the process the athletes use is one you too can acquire, and use when you're queuing in the staff canteen, browsing the supermarket aisles or looking through the cupboards at home. A process that is based upon what your body requires, not what you are used to doing. You will form good habits instead of abiding by old ones. Because you aren't simply working towards having a meal – you are building a **performance plate**.

Performance plates provide a simple solution to following the right principles at the right times. Athletes, like most performers, don't need to be weighing food or counting calories at each meal. Ensuring these principles are followed for each meal delivers what they require with minimal effort.

There are three types of performance plate, for different needs – but the reality is that, unless preparing for a specific challenge, most of us will only need to use two to support our training and lifestyle. The key is to use them at the right time within the context of your day and week.

Building Your Performance Plate

Putting together your performance plate is a four-part process. At a football club, we set up the restaurant so that the players pass each food station in the same order, adding a serving of each to their plate. This is something that can easily be replicated when building your plate at home.

1	2	3	4
Protein High quality	Carbohydrates Slow-releasing	Vegetables, fruits & healthy fats	Fluids
MAINTENANCE	FUEL	PROTECTION	HYDRATION

Maintenance, in the form of **protein,** is the first step to building every meal – end of story. We discussed back in Chapter 2 the process of your muscles constantly breaking down and rebuilding over a 24-hour period, and protein is our best aid for this.

Fuelling, in the form of low-GI **carbohydrates,** comes next, and how much is required (or whether it's required at all) depends on the training demands and your goal.

Protection in the form of **micronutrients** from vegetables and fruit, as well as healthy **fats**, is the third component of the performance plate.

Hydration is the fourth and final element; the amount required is likely to increase pre- and post-training, to prepare and rehydrate. We will discuss this in more detail later in this chapter.

There are many different types of foods that can be used to build your plate. It's firstly important to know what are the good sources for **maintenance, fuelling** and **protection** for you personally, in relation to any dietary restrictions you might have (for example if you're vegan, vegetarian or have any other

---

HANDY MEASURES – QUICK PORTION GUIDE

Building your performance plates as part of your Energy Plan should be simple. It can be hard enough to make time to cook in the first place, and the reality is that most of us don't have the extra time or the inclination to weigh our portions. So while portion weights are given throughout this chapter, for those of you who do want to be exact, here is another, quicker way to get a decent gauge on portion sizes:

- A single portion of **protein** is the equivalent of **your palm**
- A single portion of **carbohydrate** is equal to **1 cupped handful**
- A single portion of **vegetables** is **2 handfuls**
- A single portion of **fruit** is **1 handful**
- A single portion of **healthy fat** is the **size of your thumb**

If you would like more information on these different foods and weights, there are some more detailed food lists in the appendix, on page 263.

---

considerations) – and, of course, whatever your sources are, you should enjoy them.

The box opposite includes a guide to portion sizes, and there are some recipe ideas too which will give you some idea of how sources can be built into a meal – though I'm sure you will have your own ideas too.

Step 1: Maintenance (Protein) Foods

We've discussed the importance of protein as the first step for each meal, and the protein food list below includes mainly complete proteins from both animal and plant sources.

There are a few incomplete proteins which need to be combined with another incomplete protein to form a complete protein and these are marked with an*: for example, serve with basmati rice.

Remember: 1 portion = a palm

Chicken
Turkey
Beef
Eggs
Salmon
Tuna
Halibut
King prawns
Tofu
Tempeh
Quinoa
Buckwheat
Greek yoghurt (low-fat)
Beans (e.g. kidney, black, pinto)*
Lentils*
Chickpeas*

Step 2: Fuel (Carbohydrate) Foods

One of the most important things to consider with carbohydrate foods is the glycaemic index. As we said in Chapter 2, in most cases it is important to focus on lower-GI options for a slower energy release, which means keeping to wholegrain versions of your favourite staples like rice, pasta and bread. In the appendix is a full list of carbohydrates, including the higher-GI carbs, that can have a role during a heavy training programme.

Remember: 1 portion = 1 cupped handful

Oats
Muesli
Rice (wholegrain, basmati or wild)
Wholewheat pasta

Buckwheat
Quinoa
Lentils
Sweet potato
Spelt
Barley
Bulgur
Freekeh
Rye or wholegrain bread

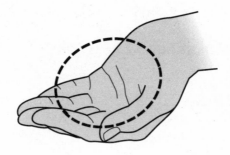

Step 3: Protection (Vegetables, Fruits and Healthy Fats) Foods

The third step in building your performance plate is adding the **protection** foods. Your need for vegetables will increase for certain meals and times (such as during the winter, which we'll discuss in the Immunity chapter, page 189), but the basic intention is to include two different types of vegetables in each meal to increase your micro-nutrient intake.

Note that 'vegetables' here means non-starchy vegetables, so things like potato and sweet potato, which contain more carbohydrate, are not included here (we are also borrowing tomatoes and avocados from the fruit category).

This structure also allows for a serving of fruit, for breakfast or snacks (e.g. antioxidant-rich berries or fibre-rich apples or pears, which also provide a valuable source of micronutrients). The overall focus is to include more vegetables than fruit within your daily Energy Plan, as they contain less sugar. Following this structure is also a great way to get you closer to that elusive five (or even seven) a day (which we'll discuss more fully in the Ageing chapter, page 241).

Remember: 1 portion = 2 handfuls

Broccoli
Spinach
Beetroot
Onion
Romaine lettuce
Avocado (half)
Rocket
Green beans
Tomatoes
Peppers
Bok choi
Asparagus
Mushrooms
Courgettes
Carrots
Peas

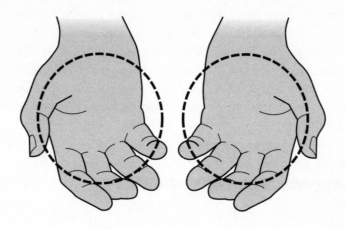

Remember: 1 portion = 1 handful

Blueberries
Blackberries
Raspberries
Apples
Pears
Kiwi
Melon
Cherries
Pomegranate
Oranges
Peaches
Passionfruit

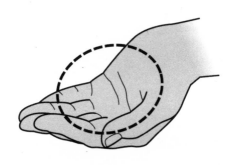

As discussed in Part I, we need to focus on healthy, mono- and polyunsaturated fats (replacing trans and saturated fats). A portion of these healthy fats should be included at each meal. A portion of oily fish (salmon, mackerel, herring, tuna, trout) as your protein food will tick this box too. As oily fish is a calorie-dense option, be careful to stick to the portion size (palm-sized).

Remember: 1 portion = 1 thumb

Extra-virgin olive oil
Rapeseed/canola oil
Seeds (e.g. chia, flax, sunflower)
Nuts (e.g. walnut, almond, macadamia, pistachio)
Avocado (half)
Oily fish

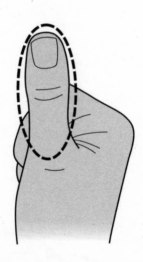

Tailoring Your Portions

Does a 100 kg rugby player need to eat more than a 50 kg gymnast? Does an 80 kg man need to eat a bigger portion than his 60 kg girlfriend? The unsurprising answer is yes, which is why in Part I we talk about grams per kilograms of body weight. The bigger you are, and the more muscle mass you have, the greater your requirement for carbohydrate, protein and fluid (the evidence on fat is less clear-cut.)

So now that we have the standard single portion sizes in place, we can look at how to adjust your size according to your weight. Rather than make things too complicated, we will use two portion sizes, **standard** and **large**. The cut-off weight is not based on gender, but on weight, and is 75 kg. So if you weigh 75 kg or less, you take a standard portion, and if you weigh more you take the larger portion, which looks like this:

Protein = Increase to 1.5 portions (**1.5 palms**)
Carbohydrates = Increase to 1.5 portions (**1.5 cupped handfuls**)
Vegetables = Increase to 1.5 portions (**3 handfuls**)
Fruits = Increase to 1.5 portions (**1.5 handfuls**)
Healthy fats = Same portion (**1 thumb**)

This isn't an exact science and you may need to do some trial and error to find the right balance to leave you sated at the end of your meals rather than not hungry or overfull; experiment and see what works for you. And if you are the 100 kg rugby player, or just training hard, you can make it a more exact science by using the tables in the appendix (see page 263) to increase your fuel and maintenance portions accordingly.

Types of Performance Plates

The performance-plate principles here are the ones we follow with our athletes, and have been designed so they can be easily applied in any setting anywhere in the world. You'll get into the habit of applying these principles every time you enter a restaurant or prepare a meal at home.

There are two types of performance plate for the average person to draw upon and structure an Energy Plan: the **fuelling** plate and the **maintenance** plate. And for those of you preparing for a major event or competition such as a triathlon, we have the **competition** plate.

The Fuelling Plate

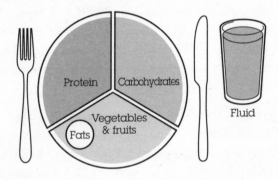

As we outlined at the end of Part I, when it comes to training, our two most important needs are pre-training **fuelling** and post-training **recovery**. The former gives us enough fuel to meet the demands of our training, and the latter helps our muscles to adapt and replenish our diminished glycogen stores. The fuelling plate is also an important tool to support energy levels during the working day, and we'll look at this in depth in the next chapter.

As you can see, the fuelling performance plate is divided into thirds to give:

1 portion of maintenance (protein)
1 portion of fuel (carbohydrate)

1 portion of protection (vegetables/fruits and healthy fats)

The fluid intake, at the side of the plate in the figure, is important with the fuelling meal, as hydration needs are greater before and after training. We'll talk about various drinks later, on page 93.

The Maintenance Plate

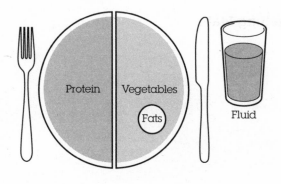

The maintenance plate is made up of:

1.5 portions of maintenance (protein)
1.5 portions of protection (vegetables)
1 portion of healthy fats

Obviously we all have different work schedules, and so we need to equip ourselves with the tools for '**flexible fuelling**', which simply means being prepared to change your meals to meet the needs of your day if plans change. For many people, it makes sense to have the maintenance meal at the end of the day when the body is sufficiently fuelled and the energy required for the evening is lowered (no training after work, for example), and carbohydrate is not required. Your fluid needs are lower here, so they can be reduced accordingly.

If the goal involves reducing body fat, there is the opportunity for flexible fuelling with your breakfast. You would use the maintenance plate if you wanted to eat *before* training (though you might want to adjust the fluid intake), unless you are training low (before breakfast), in which case you would recover using the fuelling plate after training.

The Competition Plate

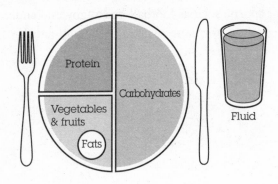

The competition plate has been designed for use with our athletes when they are fuelling up for or recovering from a particular event, such as before a football or rugby match, or an endurance event like a marathon or triathlon. It is composed of:

1 portion of maintenance (protein)
2 portions of fuel (carbohydrate)
1 portion of protection (vegetables healthy fats)

This plate is designed to deliver a larger quantity of carbohydrates to load up fuel stores in the liver and muscles. Half the plate is dedicated to carbohydrate (two portions), and extra bolt-ons like bread, juice, a sports drink or dessert can further increase the carbohydrate content. It varies a little depending on the foods, but each meal delivers at least 1 g per kilogram body weight, so 70 g of carbohydrate for a 70 kg person. This could be achieved by having a large bowl of porridge and glass of fruit juice for breakfast, or a competition plate (80 g) serving of basmati rice and a bread roll. For some competitive athletes, carbohydrate portions will need to be larger (up to 3 g per kg body weight), and the bolt-ons become essential with each meal. It can also form part of a day of 'fuelling up' involving carb-based snacks to deliver more than 6 g per kilogram body weight (we'll see this more clearly in the day planners in Chapter 6), which can be required for intermittent sports, such as football and rugby and endurance events.

Your performance plates are the foundation of your Energy Plan. They are the key to delivering an effective meal to meet your requirements on any given day. And it is critical to start with structure around these to regulate your appetite hormones, blood glucose and overall energy levels. Without having these core principles of the **type** and **timing** for each of your meals in place you might find that you are always hungry or lacking energy.

Snack Time

We've covered the three potential performance plates. So where do snacks fit in? Well, if your meals are your foundations, your snacks are the supporting structure, there to lend a hand to get you through the day. We've already established that they are important to support ongoing muscle maintenance. It's important that every snack has a function, rather than being just 'something' you blindly eat to avoid getting too hungry. Snacks play a supporting role to meet a need – but not to replace meals or cause spikes in blood glucose or unnecessary extra calories.

There are three different types of snacks, although just like with the performance plates, unless you are preparing for a competition the reality is that you will only need two:

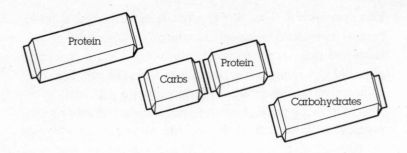

Here are some of the scenarios that may work for you. A piece of fruit has often been seen as the healthy-eating snack of choice, but there is a lot more to be gained from your Energy Plan with these performance-based snacks:

Snack 1: Protein-based (the Maintenance Snack)

- Supporting ongoing muscle growth and repair (including over-night, following hard training).
- To meet increased protein requirements alongside a training programme.
- To increase protein intake during an energy deficit (when trying to reduce body fat).
- A mid-afternoon snack to offset hunger before dinner.

Examples include seeds, nuts, low-fat Greek yoghurt, edamame or a protein shake.

Snack 2: Carbohydrate and Protein (the Training Snack)

- Pre-training snack: 1-2 hours before a training session such as mid-morning for a lunchtime session or mid-afternoon for after work.
- Post-training: a quick option to refuel and repair the muscles before your next mealtime such as finishing a class mid-afternoon.

Examples include an open sandwich with smoked salmon, or low-fat Greek or natural yoghurt with banana and nuts.

Snack 3: Carbohydrate-based (the Competition Snack)

- To top up carbohydrate levels pre-, during and post-event.
- To provide easily digested carbohydrates, which means higher-GI, close to and during competition.

Examples include home-made flapjack, banana bread, granola bar or a ripe banana.

Fluids

'What can I drink other than water?' is one of the most common questions I field. In Part I we talked about the importance of replacing fluid losses to rehydrate, and also carbohydrate while you're training – but what about day to day?

As a first principle, avoiding sugar is a good place to start. This means avoiding soft drinks, sports drinks, fruit juices and sugary cocktail mixers (see Alcohol, page 99). And as for those attractive-sounding options that list as their ingredients an array of fruits and vegetables, like smoothies and blended juices... Fruits and vegetables should be consumed **whole** where possible, to reduce sugar intake. The exception to this is if you are deliberately aiming for an increase in fruit and vegetable intake and carbohydrates. This could mean an antioxidant-rich juice when you're aiming to boost your immunity or a fruity milkshake to kick-start recovery after a gruelling training session.

When it comes to drinking fluids bear in mind that drink temperature and flavour have an effect on voluntary hydration (in other words, on how much you want to drink). An easy way to add flavour without the sugar is to infuse still or sparkling water with fruit or vegetables, as we often see in cafés now. Slices of cucumber, orange, lemon or lime all work well.

When you make a drink, make it more of an experience: use a nice glass and fill it with ice, sparkling water and a slice of lemon or lime – how much better is that experience than just filling your glass from the tap?

The Fuel Injection: Caffeine

If you've ever heard the early-morning clatter of cleats at your local coffee shop as a group of Lycra-clad cyclists come in to enjoy an espresso or two, you might have found it a heart-warming example of training being both good for our bodies and a sociable experience. Which is quite right – caffeine and coffee culture play a big part in sports like cycling, particularly with the more serious riders. And whether in the form of a fizzy drink or a morning cup of tea or coffee, they might play a big part in your own life.

Caffeine is a central-nervous system stimulant well known for its effects on alertness. It works by blocking a chemical in the body called **adenosine**, which among many other roles reduces energy and encourages sleep. Caffeine can also have potent benefits for athletic performance, such as a reduced feeling of effort during exercise.[1] This has been shown to improve endurance capacity during cycling, running and rowing and also total work-rate within intermittent sports such as football. Caffeine has been found to improve overall performance by up to 8 per cent.[2]

Caffeine has also been shown to boost cognitive performance, aiding concentration and decision-making skills, and the increase in alertness it provides can see you through periods of fatigue in the day (which we'll discuss further in Recharging, Chapter 11). It can help ensure that you turn up to meetings or social occasions on your best form.

There is a psychological factor at play too, in that if you believe caffeine will have a performance-enhancing effect, it will affect how you will respond to that cup of coffee.[3]

Its effects, then, are potent – it's perfectly legal, but none-theless a performance-enhancing drug. But it's effective only in the right quantities, which differ from person to person.

The Caffeine Dose

At a football club, many players generally take between 1 and 3 mg per kg of body weight of caffeine around 45–60 minutes before kick-off, depending on their individual tolerance, so that it peaks in their blood for the match. So a 70 kg player would take between 70 and 210 mg of caffeine.

To give you an idea of what these quantities look like in real terms, a can of Red Bull and a single espresso both provide about 80 mg. A teaspoon cup of instant coffee pro-vides about 55–70 mg and a cup of tea has around 25–45 mg. A double-shot flat white or latte in a coffee-shop – in other words a drink usually based on a double espresso – would offer around 150 mg of caffeine.

Most athletes across many different sports will still have a cup of coffee pre-competition, but for consumption just before or during competition (such as out on the track, on the bike or at half-time in team sports) there are now more efficient delivery options available in the form of supplements and sports nutrition products. One novel product is caffeinated gum, which can contain as much as 100 mg caffeine per stick and was first used by the US military. The caffeine is absorbed more quickly through the membranes in the mouth than it would be through the stomach via a drink.[4]

Caffeine shots, chews or gels, which are carbohydrate-based, are popular in both endurance and team sports as they also support the carbohydrate targets for refuelling during competition (which we discussed in Chapter 3). The caffeine dosage varies from around 50 mg to 150 mg (so from around the same as a capsule-delivered coffee to that of a double espresso).

Finally, there is the no-frills delivery system of a caffeine tablet, which gives 50 mg. Many athletes still prefer this approach, due to the carbohydrate products causing GI issues during competition.

Whichever method an athlete uses, it will be trialled during training first, never introduced in a big competition; principles to consider for yourself that we'll come on to shortly.

Should I Use It?

One of the negatives around caffeine has always been that it was believed to be a diuretic, but now there is evidence to suggest that this isn't the case, at least not for moderate coffee drinkers.[5] But there are, of course, other drawbacks to caffeine. Some people get jittery after coffee and other caffeinated drinks, some can't sleep, some crash terribly when it wears off – although others appear to be able to drink it regularly through the day and evening with no repercussions. There are multiple reasons for this and the research is starting to highlight different genes causing people to be fast or slow metabolisers – but it is still very early days in our understanding of personalising caffeine intakes.

Caffeine can cause GI upsets, elevate the heart rate, induce anxiety and restlessness, irritability and headaches. It can remain in your system far longer than you might think (it has a half-life* of three to five hours in your blood), disrupting sleep and having knock-on effects on your mood and cognitive capabilities the next day. You may be familiar with that 'caffeine hangover' feeling following a day of coffee excess when you've been chasing a deadline or studying late. Used in the wrong quantities, its stimulating effect on your central nervous system is a double-edged sword. Caffeine can be a performance-diminishing drug,

*The time it takes for your body to process half of the original amount.

depending on your personal tolerance and the strategy you use around it.

If you are particularly sensitive to caffeine, the chances are you already steer clear of it. There are many athletes who are sensitive to caffeine's effects and don't drink coffee during the day or use it as part of a training or competition strategy. If you do want to avoid it, focus on your fuelling instead to maintain your energy levels.

For some people, caffeine can become a fuel they think they need to get through the day – they need their morning cup to get going – but as we've seen from the potential side effects, it isn't necessarily a sustainable fuel for your health or performance. If you structure your food intake in a way that supports energy levels throughout the day as part of your Energy Plan, caffeine can simply be something that you can choose to use to give you that fuel injection when you need it.

You can integrate caffeine into your Energy Plan following the principles below, just as with **type, timing,** and **amount** of food – the TTA model, as mentioned in Chapters 3 and 4.

TYPE

Trying out different coffees at coffee shops, and different modes of delivery when making at home (such as capsule machines, which deliver around 60 mg per capsule), as well as buying and using different energy drinks and caffeine supplements, will allow you to see which gives you the best response.

It's worth noting that a capsule machine has the advantage of delivering pretty much the same amount of caffeine every time, while high-street coffee can vary dramatically depending on the serving size and preparation, ranging from around 70 mg to as high as 387 mg per cup (for a Starbucks Venti filter coffee).[6]

TIMING

For optimal performance in an event or exercise session, caffeine should be taken around 45 to 60 minutes beforehand[7] for levels to peak in the blood. But the half-life of caffeine in the blood is 3–5 hours[8], depending on the person, which can mean that your late-afternoon coffee may also serve as a useful pick-me-up before the gym. But do bear in mind that caffeine late at night can disrupt sleep, so depending on your personal tolerance to it, you might need to try to avoid caffeinated products after 4pm.

Try to use caffeine strategically, not habitually during the day. It may be that a large dose of caffeine from your high-street café chain isn't always required; you might get equally effective results from something with a lower caffeine content, such as English Breakfast or other kinds of caffeinated tea. And be alert to how the effects vary depending on the times of day you consume caffeine and the variety of your activity.

AMOUNT

Guidelines for athletic performance state that the optimal dose starts at 1–3 mg per kg body weight (70–210 mg for a 70 kg person). This can increase to a very large dose of 6 mg per kg body mass in elite athletes; in some cases, depending on body weight, this is above the recommended safe daily amount for that same 70 kg athlete. But practically it is best to start low (perhaps with a single cup of coffee) at the right time before training or an event.

While there is no set or single daily caffeine limit, a maximum of 400 mg per day is recommended and generally considered to be safe for healthy adults (a double-shot flat white contains around 150 mg). The main exceptions to the rule are pregnant and breastfeeding women, for whom a maximum of 200 mg per day is advised.

WHAT CAN WE LEARN ABOUT CAFFEINE FROM THE US MILITARY?

Science is starting to investigate a much more personalised way of using caffeine. A 2018 US Army research centre conducted a study on an algorithm of its own creation to identify the most effective caffeine-dosing strategies to counter sleep loss.[9] It found that the algorithm identified strategies using smaller doses at specific times enhanced cognitive performance (such as reaction times) by up to 64 per cent. Reducing caffeine consumption by up to 65 per cent achieved equivalent improvements in alertness to the other studies it was compared against.

The algorithm is to be made available to the public through an open-access web tool.[10] Do be aware though that this is the first study of its kind, so it's early days yet, but it offers some exciting possibilities for the future of caffeine dosing.

When planning your caffeine dose, remember that a 4pm pick-me-up coffee at work might be enough to deliver the requisite boost to performance in your evening training session at 6pm. This could prevent you 'double-dosing' – having another coffee pre-training that might add little to performance, but would certainly add to your caffeine consumption.

Rocket Fuel: Alcohol

In a section devoted entirely to drinks, it would be remiss not to talk about alcohol as part of your Energy Plan. The good news is that you can still enjoy a few off-Plan rewards in the form of drinking alcohol sometimes. But it requires a little thought and planning if you don't want to derail things too badly.

While alcohol has the capacity to leave you well-oiled, it certainly doesn't do your engine any good. Alcohol is energy-dense (containing 7 calories per gram) and nutrient-poor, hence the popular expression to describe it – 'empty calories'. Combining alcohol with sugary mixer drinks only makes things worse, of course, and if we're running an energy surplus regularly through drinking, the result will be fat storage. It's not called a beer belly for nothing.

How does it affect your Energy Plan?

Alcohol interferes with how the body produces energy. Your liver has to work overtime breaking it down – it takes on average an hour to remove one unit of alcohol (about half a pint of average-strength beer or half a 175 ml glass of wine) from your system, depending on factors like gender, age and weight.

This means the liver is less efficient in producing and regulating blood glucose, and low blood glucose inhibits your ability to maintain intensity during hard training sessions.

Drink	Calories
Beer (pint)	215
Wine (standard glass – 175 ml)	159
Vodka (one shot)	54
Whisky (one shot)	64
High-calorie mixers: cola, cranberry or orange juice *Low-calorie mixers: diet tonic, diet cola or soda water*	

Drinking alcohol after training can also interfere with muscle protein synthesis and recovery.[11] We know that your body needs carbohydrate, protein and fluid after training, so hitting the bar instead of having your recovery snack or meal can affect your muscles' refuelling and repair, and leave your body dehydrated. Sleep is also affected, compromising your recovery further.

How to Do It

You can still enjoy a drink (within UK government guidelines) as part of your Energy Plan, but there are a couple of principles to adhere to.

Firstly, plan a night out so it won't affect any key training days or fall straight after you've been training or performing hard. Your body is still adapting more than 24 hours after a hard training session, so give it a chance to make those gains.

If there *is* a night out scheduled straight after your workout or match, get your recovery done first – rehydrate and have a meal containing carbohydrate and protein to start to refuel and repair muscles before having your first drink.

Alcohol is, of course, just one of many temptations and potential roadblocks that might threaten to bring your Energy Plan grinding to a halt. It's important to remember that there are going to be moments when you choose to relax your Plan – and moments when it's not your choice and things just don't go to plan. We'll explore these further in the Winning Behaviours chapter on page 115.

CHAPTER 6

24/7 Fuel: The Planners

Within any professional sport, training is organised into cycles with each cycle playing a part in preparing for the competition. Training and nutrition are engineered to allow the athlete to hit peak form for each competition. In many sports these competitions are far apart (for example the Olympic Games), but in professional football, where the games come thick and fast, the organisation of training runs week to week. This is called a **microcycle**. During the season a microcycle would typically include one or two matches, a recovery session following each match (which may involve pool work, massage or cryotherapy), light training, hard training (on the pitch), resistance training in the gym and a rest day. This would all fit into a weekly training planner.

With this planner in place, my role is then to devise the team's nutrition strategy to not only meet but amplify the desired outcomes from training, whether that's increasing strength or endurance, reducing body fat, increasing immunity (which we'll discuss in more depth in Chapter 12) or enhancing recovery.

Most of us probably understand structure in our own lives as being week to week. So effectively we have our own microcycle within which to engineer our Energy Plan. Let's start with planning our days first and then build up to planning in weeks.

24-Hour Planners

With the raw materials for your Energy Plan in place, it's now time to start pulling it all together as part of a 24-hour strategy to deal with your demands. As ever, it's important to remember the TTA model: **type, timing, amount** (see page 62).

Bear all of these in mind as we now look at the three broad classifications of day we will use with the Energy Plan, depending on your goal and the physical demands for that day.

1. Medium Day

- **Two fuelling meals.**
- **One maintenance meal.**
- **Two snacks (one fuelling, one maintenance).**

What's it for:

- Starting point to support general fitness
- Single-session training day

Your medium day is designed to allow you to fuel and recover from daily demands that would typically involve one training session. The example below assumes you are training in the morning; for evening training make the breakfast a maintenance meal and use the fuelling performance plate for dinner.

AM Training

The flexible worker: The ideal structure when starting a new training programme is to have a fuelling breakfast before training and a fuelling snack to recover (if not having lunch until later), or taking lunch as your recovery (ideal for those with flexible working hours).

The 9–5er: of course, not everyone works flexible hours. So let's talk about training before breakfast. This takes a

The flexible worker	
Timing	Feeding
Breakfast 8-9am	Fuelling
TRAINING	
AM Snack	Fuelling or Maintenance
Lunch 12-2pm	Fuelling
PM Snack	Fuelling or Maintenance
Dinner 7-9pm	Maintenance

little more problem-solving: a straightforward option is to train before breakfast and use the fuelling meal at breakfast for recovery but the downside is that for hard, intense sessions, you could be short on energy. In this case there are two options:

1. Have a fuelling snack before training
2. Have a fuelling meal for dinner the night before, to increase your fuel stores ahead of this training session.

If you are training low during the week, it is important that weekend sessions are fuelled. This is so your body remains accustomed to using carbohydrate as a fuel (otherwise carbohydrate metabolism can become less effective); plus, for prolonged weight-bearing (e.g. marathon training), this is important as it will reduce the stress on your bones.

The 9–5er	
Timing	Feeding
TRAINING	
Breakfast 8–9am	Fuelling
AM Snack	Fuelling or Maintenance
Lunch 12–2pm	Fuelling
PM Snack	Fuelling or Maintenance
Dinner 7–9pm	Maintenance

PM TRAINING

After-work club: When training in the evening the priority is to make sure the muscles are fuelled before training and recovered after (especially if it is a hard training session). If it's a hard session, you need to ignore the old wives' tales about eating carbs late, as you will need them to refuel your muscles. The exception is if it is a light training session.

You will also see that as you are increasing your fuelling later in the day the maintenance meal is now taken for breakfast and, if a morning snack is required, it should be a maintenance snack.

Finally, while we are on snacks – each of the planners contains two snacks to increase the options around training. But if you prefer to take just one, that's fine too.

This medium day is the go-to option for your Energy Plan. It's a great all-rounder. If you do not already have a nutrition

After-work club	
Timing	**Feeding**
Breakfast 8-9am	Maintenance
AM Snack	Fuelling or Maintenance
Lunch 12-2pm	Fuelling
PM Snack	Fuelling or Maintenance
TRAINING	
Dinner 7-9pm	Fuelling or Maintenance

structure to your days, this is the place to start. For those of you with a training plan already in place, this is still the place to start.

While it is designed with a training session in mind, if you are coming to the Energy Plan without any exercise or nutrition structure in place and you are simply at the beginning of your journey, you should first establish your goal, as per Chapter 4, get training and then look to adopt the medium day. Of course you could tailor your nutrition before starting to train, but in my experience it's best to just get started and throw yourself into the training – your body will then be screaming at you to fuel it effectively and you'll definitely notice the difference. Often it's too easy to procrastinate, thinking you'll wait and do the tailoring once everything is perfect.

Anything that marks the beginnings of new, sustainable habits is a change for the better and, as I hope the Energy Plan is something you will still be using years from now, better to start slow and make gradual changes than attempt to get your engine from zero to sixty too fast and find yourself crashing and burning.

Please be aware that it will take your body a while to adjust to this new pattern of eating. You may find that your appetite hormones adjust and you are most hungry in the evening – this is normal. What we are planning to ensure is that you are not hungry or fatigued when you need the energy – that's to say, during training or the working day.

If you haven't up until now planned your nutrition as carefully as your training, it can be a big transition to mapping out your fuel for individual days (and weeks, which we'll come to). Start slowly with the options described above for the first month, after that you can use your monitoring tools to reflect and refine – see Chapter 8.

2. Low Day

- **One fuelling meal.**
- **Two maintenance meals.**
- **Two snacks (both maintenance).**

Low day	
Timing	Feeding
Breakfast 8-9am	Maintenance
AM Snack	Maintenance
Lunch 12-2pm	Fuelling
PM Snack	Maintenance
Dinner 7-9pm	Maintenance

What's it for?

- Reducing body fat
- Rest day
- Travel day

The low day is of particular use when a reduction in body fat is being targeted; when that is your goal you will include more of these in your week. The reduction of carbohydrate outside of lunchtime is a crucial part of this strategy, and taking on your carbs in the middle of the day will ensure you maintain some fuel during the day when activity levels are typically higher than the evening, when less fuel is required.

Your rest days, when you don't do any exercise outside of general physical activity, should be low days, as they include sufficient amounts of protein for your body's maintenance and less carbohydrate that won't get used, while days you spend travelling (see Chapter 13) should also be low days, as you're likely to be less active. Those who have adapted to the use of medium days can start to integrate low days if they fit with their goals.

When training on a low day as part of a strategy to reduce body fat, it is important to experiment with the timing of your fuelling meal. Many athletes would usually aim to train low following a protein-based breakfast and use the fuelling lunch to recover. But you will need to trial what works best for you.

When you start to introduce more low days into your programme it becomes vital to keep up with your wellness monitoring, which we'll come to in Chapter 8, to see how your nutrition is affecting mood, sleep and energy levels – as these can all be negatively affected by low energy intakes.

3. High Day

- **Three fuelling meals.**
- **Two or three snacks (fuelling).**

What's it for?

- Increasing muscle mass
- Weight gain
- Double training day

High day	
Timing	Feeding
Breakfast 8-9am	Fuelling
AM Snack	Fuelling
Lunch 12-2pm	Fuelling
PM Snack	Fuelling
Dinner 7-9pm	Fuelling
Snack	Fuelling/Maintenance (optional)

At the beginning of the previous chapter we looked at the boxer, the footballer and the sprinter, and this is the kind of day all three will incorporate into their Energy Plan.

If you're serious enough about your goals to be doing double-session training days, then there's no point even attempting them if you're underfuelled. Incorporating high days when you need the energy most will make sure you can get the best out of your training and begin to recover effectively. If you are eating dinner early, adding in a protein-based evening snack will help to support overnight muscle protein synthesis (MPS).

If preparing for a challenge or competition – such as a marathon, triathlon or a football or rugby match – the **competition plate** (see page 90) should be used to increase daily carbohydrate intakes (for our elite athletes this can be over 6 g per kg body weight). Fuelling days, containing competition plates to increase carbohydrate intake, are typically used in the 24 hours before and after competition in football and other team sports, but can extend to 48 hours for endurance events, such as preparing for a marathon. This is the structure we use for our marathon fuelling plans with *BBC GoodFood* and is used on the day before competition, on competition day and the day after to recover fuel stores.

The Microcycle: Your Weekly Planner

The weekly planner brings everything together to deal with the week ahead as part of your Energy Plan.

Here is a case study of one of my clients, Mia, who has been using a week planner. Use her planner on the next page as the basis for your own.

Training: Mia has only recently started a training programme. She isn't particularly sporty, but is keen to overcome that to make some changes to her life. Her Energy Plan is aimed at building fitness and reducing body fat. Mia likes to train in the morning, and on non-training days she has also been trying to improve her activity levels during the working day – her step count is up significantly.

Day planners: You will see that her week consists of five medium days and two low days. She's recently added the low

	Mon	Tues	Wed	Thurs	Fri	Sat	Sun
Breakfast	M	F	F	M	M	F	M
AM TRAINING	HIIT Class		Weights			Long run	
Snack (Recovery)	M	M	M	M	M	F	M
Lunch	F	F	F	F	F	F	F
Snack	M	F	F	M	M	M	M
Dinner	M	M	M	M	F	M	F

Mia's week planner

Key:

M = Maintenance meal or snack

F = Fuelling meal or snack

days, after spending the first eight weeks using the medium-day structure. She's now ready to take the next step.

Mia likes to have a low day on a Monday, to get back on track after a relaxed weekend (Sunday evening is always a big, chaotic meal with her family). She prefers to take her fuelling snack mid-afternoon on Tuesday and Wednesday, as these are particularly long days at work, as otherwise she is extremely hungry when she gets home for dinner.

Friday night for Mia usually involves going out with her partner or friends, so she schedules that as a fuelling meal, ahead of her longest and hardest session of the week on Saturday morning (after a fuelling breakfast). Saturday is also the only time she has carbohydrate in her recovery snack – her favourite is a home-made smoothie with milk, banana and mixed berries – so she has plenty of energy for the rest of the day with her family.

The aim is for you to plan your week during your **weekly check-in** (covered in the chapter on monitoring, page 127). The planners can be flexible, and depending on your progress, you can amend how many different types of day you need in your week:

- Start using **medium days**.
- Add **high** days if your training volume demands it.
- Add **low** days to accelerate your losses in body fat.

I can't stress enough that, whatever your goal, monitoring your progress during the weekly check-in is key to assessing progress and then refining your plan.

CHAPTER 7

Winning Behaviours

At the Olympic village in London during the 2012 Games, the food hall was producing something like 60,000 meals per day. And when I say 'hall', what I mean is something more akin to an aircraft hangar, a sensory overload of an operation filled with restaurants offering food from around the world to cater for the different needs of different continents – something for someone from anywhere in the world to eat.

A curious mix of adrenaline, nerves and excitement bubbles around an Olympic village during the Games. There's the unburdened and unbridled joy of those who have already competed and triumphed, rubbing shoulders with the much more reserved masks of concentration and apprehension worn by those who have yet to compete.

The stars of track and field and cycling will be using their Energy Plan to ensure they are fuelled for their coming events, while the boxers, judoka (judo players) and divers will be using their own Energy Plans to manage their weight. These are dedicated athletes who have spent the last four years training for this one goal. They have almost certainly spent their whole lives dreaming about it.

Which is why it might sound strange to you to learn that, right in the middle of the food hall was a McDonald's. This is the case at every Olympic Games and it always commands a lot of media attention. This kind of temptation at the door is just one of a number of things the athletes have to consider when managing their environment.

Managing Your Environment

Once you have decided on your goal, but before you dive into the detail of your performance plates and snacks, it's vital to have the right infrastructure in place. Professional athletes obviously have a lot of help with this thanks to the support staff at their disposal, but with some management of your own environment you can build your own performance infrastructure and create your own **winning behaviours** to help you reach your goals with the Energy Plan.

With any person I work with, whether professional athlete, performing artist or corporate client, I want to know as much as possible about how their lives are set up so that their Energy Plan can be set up to provide a lasting solution. I work by dividing life set-up into four sections:

1. **Home** When starting out on the Energy Plan, a revamp of the home environment is vital. Ensure that all the old food, drink and temptations that aren't part of meeting your goal are thrown away or donated and that the cupboards are well stocked with the shopping list in Chapter 10.

    Introduce some little nudges into your home environment, whether that's colour-coded reminders on your smartphone for your high, low and medium days or always leaving your packed gym bag by the front door the night before you have training. You need to make it as easy as possible to make the right decisions for your Energy Plan.

2. **Work** The workplace is the blind spot for many people, and often requires a bit more effort than establishing good habits at home. Ask yourself:

- Do I have the utensils I need at work?
- Are there snacks available from the snack list in Chapter 5?
- Do I eat lunch at my desk?

If you're like 62 per cent of American office workers, then you might answer yes to the last question, but this is something that isn't negotiable: **eating at your desk runs counter to all of the advice in this book**. Get up and have your lunch elsewhere. Not only the obvious lack of physical activity during the main opportunity in the day, but also the time to switch off, socially engage with colleagues or even just give your brain a break with some new stimulus. And with the snacks and utensils, if there aren't any at work then you'll need to bring them in yourself; this will make sticking to your Energy Plan so much easier.

3. **Training** You would be amazed by the number of people I meet who train on the other side of town or somewhere miles from where they live or work because they like a particular personal trainer or gym. But in my experience a fantastic gym in the wrong part of town reduces compliance with your Energy Plan. Find somewhere close to home or work so that, when time is of the essence, it doesn't so easily become something you don't have time for.

   And don't forget the 'f' word – make your sessions as **fun** as possible. During the week, when time can be tight, it might be difficult to get anything other than your necessary training dose in, but come the weekend, the grunt work is done and you can change things up. Get out into the countryside on your bike, meet friends for a class – whatever it takes to make you feel like your training is an enjoyable and engaging part of your free time, rather than something you need to get through before enjoying the weekend.

4. **Significant others** Get your significant others on board and you have every chance of succeeding in your Energy Plan.

Get this wrong and it could be a struggle. Your partner, wife, husband, girlfriend, boyfriend, housemate, family members, best friend, workmate or whoever are your closest people need to be 'on plan' and supporting you, which at times might mean plenty of encouragement and at other times might mean some of the tougher varieties of love.

Help them to help you. Make sure they really understand your goal; use some of the language you used when setting it to make it feel tangible and important to you. Share your motivation and how it makes you *feel*. Explain that this isn't a short-term diet, but rather a long-term lifestyle change.

If they're really supportive, or perhaps become inspired by your approach, they might get started on their own Energy Plan too, so that you're doing it together to reach your own goals. This is like creating your own team environment, with everyone working together towards a common purpose. And within your own team environment you can start doing things like taking it in turns to try cooking new dishes, offering support to one another and putting your results, comments and progress up on the fridge or on a shared app.

Peer support can be crucial to inspire you both towards meeting your goals with both training and nutrition,[1] and being on a journey can really help push you to get on with your Energy Plan on a cold November morning. And this support doesn't necessarily have to come from someone you eat or train with every day. It could be a colleague or friend you check in with once a week to help keep you both on track.

And finally, sad as it is to mention this, any performance environment, from an elite sports team to businesses, will contain negative people with unhelpful comments about what you're working towards. So learn to filter this out – or to use it to spur you on – and, fairly obviously, make sure you don't have this kind of person in your support team.

Eating at Home

The home environment is the one in which we're able to exercise the most control. Of course you already know how to eat, and you probably do most of your eating at home – but how can you build your home environment to help you cook and eat in a way that better supports your Energy Plan? Here are a few golden rules to enforce at mealtimes:

Don't Sit Down at an 8 Out of 10

We have already discussed, in Chapter 3, the importance of meal timings to regulate blood glucose for energy levels and to keep appetite in check. Your mid-afternoon snack is crucial to offsetting hunger at dinner time, so that you're in control and not tempted to overeat.

In Chapter 8 we'll look at tracking – tracking appetite is another tool at your disposal (out of 10) you'll learn that a rating of 8 or more is in the danger zone at mealtimes. Look at how your day is structured, the timing of and fuels that make up your lunch and afternoon snack, and look to get them aligned so that, when dinner time comes, you're not sitting down at an 8 out of 10.

Slow It Down

Sip a glass of water with your meal. Chew your food thoroughly. Most importantly, *take your time*, so that your hunger hormones can signal to your brain that you're satisfied; this can take around 20 minutes from eating to registering.

Satisfied, Not Full

There is a Japanese practice called *hara hachi bu*, in which people eat until they are 80 per cent full. So get to know what this feels like to you. Our instinct can often be to clear our plate, but if we engage more fully with how we *feel* as we eat – if we're more mindful about our eating – we can learn to eat until we are satisfied, not full.

Minimise Distractions

Sitting at the table to eat, with the television off and no smart devices at hand, might sound like the kind of advice a parent might insist on (or at least try to) with their children, but it is important to be aware of how these external stimuli affect our eating behaviours. I'm not about to suggest you live like a monk at mealtimes, but having the television on can affect the amount of food we eat.[2]

The work of Professor Charles Spencer's Crossmodal Research Laboratory at Oxford University, which studies how we integrate stimuli across our senses, has found that music with a faster BPM (more beats per minute) makes us eat faster.[3] So if you want to listen to music with dinner, best to opt for something slower and leave the party music for later.

And then, of course, there are our smartphones, those little devices tailor-made for taking snaps of picture-perfect plates and uploading them to Instagram so we can revel in the dopamine high when the 'likes' roll in. At the sporting institutions I work in, phones are banned from the table so that the focus is on spending time with teammates. To be fully present at mealtimes, whether with your partner, your family or friends or eating alone, phones should be off the table before eating. If you really must, get an initial snap of your plate – but then put your phone away and don't post the picture till later, otherwise you'll only want to check your phone throughout your meal.

Your Home Environment

After a long day at work and then preparing some food, just slinging dinner down on the table and tucking in isn't giving you the full experience you've earned. Setting your environment up at mealtimes can vastly improve your home form, and can make even a quickly made meal, eaten alone, feel like more of an occasion rather than purely functional.

Back in 2007, when I was working with the British athletics team, we had a training camp in Macau, China, prior to the World Championships in Osaka. Athletes like Mo Farah, Jess Ennis-Hill, Phillips Idowu and Kelly Sotherton were all cooped up together for up to three weeks, and, as is quite usual in situations like this, nothing much changed around the hotel and so the focus came to be around food.

This close to a major championship, the athletes' training is lighter – the intensity, to recreate the demands of the competition, remains high, but the volume of work reduces – so the fuelling requirements are less. So, in an environment where food had become the focus but the fuelling needed to reduce, we made a slight but significant intervention: we used smaller plates to make portions look bigger. We habitually expect our plates to be full and, as strange as it might sound, recent research looking at self-serving food sources showed that eating from a bigger plate increased the amount self-served by 41 per cent, therefore increasing the likelihood of more being consumed.[4]

Food is fuel but it's also to be enjoyed, and having some key homeware ingredients to go with the edible ingredients that make up your performance plates can make all the difference:

- Small plates
- Heavy bowls
- Heavy glassware and cutlery (this affects how we perceive the food we are eating; in a restaurant setting it would make you pay more for a meal,[5] so why not at home?)
- Good-quality but low-maintenance tablecloth, place-mats or coasters – or any other decorative items you prefer – to finish things off, aesthetically speaking.

Eating Out

So, you've established your home environment, but now comes the big test: eating when away from home.

Whether you're an Olympian or a business person, eating out is an important event. Time spent sharing a meal with colleagues, family or friends is a big part of our culture, and for athletes this is often the only time, other than in the training environment, when all the coaches, support staff and performers are together in the same room. It's an invaluable time for people to bond.

In the workplace, eating out is just as important. Many meetings are conducted over dinners and lunches, and the success or failure of these can ride on creating the right impression, atmosphere and overall experience at the restaurant table.

And when eating out socially, you want to relax and have a good time, of course, but the stakes can be just as high to deliver a performance: maybe you have a dinner date with a potential partner where you're anxious to deliver a good impression, or you're at a loved one's birthday dinner that you want to make special for them.

No matter what the occasion, there will be plenty of temptation on the menu. So, how to manage your Energy Plan in these more challenging environments?

Get on the Front Foot

Have you ever noticed, when you're part of a group in a restaurant, how easily the person who speaks up at the very beginning (maybe to answer the question, 'How many courses are we having?') can set the tone for the whole meal?

How hard would it be for you to be proactive and set the tone for the meal? Depending on how close your friends are, you should be prepared to take some stick from them, so deal them in: tell them you are working on plan, towards a goal that you are serious about. And no, just because you are tailoring what you are eating, *it does not mean you are on a diet*.

Build Your Performance Plate

When you're reading the menu in a restaurant, apply the principles of the performance plates we set out in Chapter 5. Look first for where your maintenance hit (protein) is coming from, and then build from there depending on what your needs are for that meal. Be prepared for the possibility that you may need to build your own performance plate, choosing from separate parts of the menu; this might involve ordering a few sides and a starter as your meal, rather than simply choosing a main course. Unconventional, maybe, but doing it gets you on the front foot at the beginning of the meal and primes you, which is essential – because you will need some confidence in your Energy Plan and yourself to do what comes next.

Don't Be So British!

As well as being prepared to go off-piste, as detailed above, you might need to ask questions about what's in your food. It's time to be a little less reserved about things (a common affliction among Brits), and to remember that you're not offending anyone – you're engaging with your food choices and your Energy Plan.

The classics to look out for:

- If this is a maintenance meal you need to say a firm no to fries or potato of any kind. Remember, **if it comes to the table you will pick at it**.
- **Limit the condiments** - ketchup, soy sauce, mayonnaise and mustard are all calorie-dense.
- Ask for **sauces and dressings on the side** – otherwise your eggs Benedict could be swimming in hollandaise.
- **Grilled or steamed** options beat fried every time.

- Don't be shy about asking for a **separate side salad** or **vegetables** to make up the protection portion of your meal – you can go without another glass of wine if you don't want to add to the cost.

The Big Event

Whether you have an important work lunch, are celebrating a loved one's birthday or are enjoying a meal as part of a weekend away, do your homework in advance, especially if you're the one organising it. Knowing the menu in advance can help you feel in control before you arrive, especially if there's a lot riding on the meal. In all the above scenarios, be prepared, if all else fails, to have a plan B, which we cover next.

When the Wheels Come Off: Restructuring Your Energy Plan

Having a backup plan might mean taking your maintenance meals at another time of day, so that when you're eating in a restaurant you can use your fuelling option. Or, if you're meeting a client in an Italian restaurant famed for its pasta or your friend's birthday is at their favourite burger place, be prepared to live a little and use flexible fuelling, as discussed earlier.

And then there's plan C. Sometimes things just don't go according to plan, and you will have a 'bad' day (although it might feel more like a really good day!), whether that involves too many drinks with friends, bingeing at the biscuit barrel or consuming what felt like double your normal daily food and drink intake. Rather than beating yourself up about it, enjoy it, but do have a contingency in your back pocket to get you back on track as soon as you can, so that one meal or night out doesn't derail you. This might mean a pre-breakfast training session the next day, when your muscles are well stocked with fuel from dinner the night before and you're in the perfect position to burn off some fat. Admittedly, you probably won't feel like this the morning after too many drinks, so if this is the scenario, give yourself a break the morning after and just get yourself back on track as soon as you can. Combine this with a **low day,** as discussed in Chapter 6, and you will feel yourself taking back control.

You can see this as a damage-limitation strategy; just as many of the people I work with do, so that you're free occasionally to enjoy a meal or other event that breaks with your plan. If the wheels really do come off, you know that you have the tools and confidence to apply the **winning behaviours** that many of us use to get our Energy Plans back on track.

Moving the Needle:
Monitoring Your Progress

Having invested time and effort in tailoring your eating and training around meeting your goals, you'll soon want to know that you're moving the dial in the right direction with your Energy Plan.

Within elite sport we have the ability to measure all sorts of variables to see how athletes are coping with training and optimising performance, from GPS units to measure footballers' sprints, accelerations and decelerations during training to biomechanics experts looking at long- and triple-jumpers' technique and trajectory. We can measure a host of blood markers such as hormones, immunity and inflammation, and carry out many other physiological tests. But should we be doing all this?

Such is the pressure in professional sport to deliver winning performances that we're always looking for the next big thing to help give a competitive edge. This thirst for data and technology is replicated in the public domain; it is estimated that by 2020, there will be 500 million wearable fitness devices being worn.[1]

But how much information do we really need? In professional sport, we've learned that measuring too many things can muddy the water around how the athlete is coping with training. It's too much information. Just because you *can* measure everything doesn't mean that you *should*. It comes back to the idea discussed at the beginning of Part II: Why the Energy Plan?

Why are you measuring it? Is it essential? What question are you trying to answer?

Too Much Information

Let's return to our boxer at the start of Chapter 4. He needed to reduce body fat in the weeks leading into a fight to make his weight category, a process common in combat sports such as mixed martial arts (MMA) and judo, as well as horse racing. In combat sports a competitive advantage can be gained from training at a heavier weight and then reducing it as the fight approaches.

In the boxer's case, what he needs to measure is simple, right? – it's his weight. Well, it's actually not quite as simple as that. Although the focus is on weight, his body composition will be regularly monitored to ensure he is losing the right type of weight – reducing his body fat while retaining muscle mass. This is before there is even any conversation around dehydrating (a tactic commonly used to 'make weight').

A Weighty Issue

This leads us on to the big issue. Overweight, underweight, I need to lose some weight, why does my weight keep fluctuating...? Weight is a heavy word, from which there is no escape in the world of nutrition. But is weight really a good measure of progress when you're trying to adopt the Energy Plan?

The first place to start is with the age-old rhetoric that 'muscle weighs more than fat'. In fact, a kilo of muscle weighs just the same as a kilo of fat – and, for that matter, as a kilo of potatoes. What is true, however, is that muscle is more dense than fat, as it contains fibres that contract to allow us to move. So in effect, this means that the equivalent weight of muscle will take up less space than fat: it weighs more *in relation to volume.*

So if you're starting a programme of increased resistance exercise and improved nutrition as part of your Energy Plan and you're using weight as your measure, you might find yourself disappointed by the results. You might find that your weight hasn't changed much at all, but what might be happening is that your muscle mass is increasing and your body fat reducing. This would happen without there being much change in your weight.

Increasing your muscle mass should, eventually, help get you closer to a more athletic, toned physique. And increasing muscle mass has the double benefit of increasing your capacity to burn more fat, as muscle tissue is more metabolically active than fat, with containing those power generators we introduced in Chapter 1, mitochondria. So the sooner you start incorporating some resistance training to increase your muscle mass, the better you will look and feel, and the better your body will be able to burn fat. It's a snowball effect – and if you're simply going by your weight, particularly at the beginning, you might find yourself getting discouraged before you've even started in earnest.

BMI

So if weight's coming up a bit light in its effectiveness, what about that more scientific-sounding stalwart the Body Mass Index (BMI)? The groundwork on BMI was developed in the 1830s by Adolphe Quetelet, a Belgian statistician and sociologist, although the term wasn't coined until the 1970s. BMI is used to determine the body fat content in an individual – or broader populations – by the following calculation:

$$\text{BMI} = \frac{\text{Weight (kg)}}{\text{Height (m)}^2}$$

An individual can then be categorised as being under-, normal, overweight or obese. A BMI of 25 or more is overweight, while the healthy range is 18.5 to 24.9. So, a man who is 1.83 metres tall (six foot) and weighs 86 kilograms (around 13 and a half stone) would have a BMI of 25.7, which is classified as 'overweight'.

All of which might sound fine until you realise that the BMI isn't really assessing the body's fat content at all, and is simply falling into the same trap as those of us gauging progress solely on weight, albeit in this case in proportion to height. A six-foot-tall professional athlete weighing 86 kg would have exactly the same BMI as his couch-potato counterpart. Does that mean the athlete is overweight? BMI is effective enough as a blunt instrument to help health professionals make an initial assessment and for measuring general populations, but for the individual person wanting to assess their body and its changes, it falls a long way short. Our lifestyles have changed a lot since the 1970s, thanks to the prevalence of gyms and our approach to exercise and nutrition, and the BMI feels now like an old-fashioned tool for our needs.

So let's look at the methods that are more useful to monitor our progress. With each method I have provided both a **basic** and an **advanced** option for you to concentrate on depending on your means and how far you want to push things. Anything that involves greater resources, such as a private clinic or a greater investment in your time, will be an advanced option, and there is a suitably budgeted basic option too, which can be just as effective if used correctly. This is about keeping the burden as low as possible – because we all have busy lives – while also ensuring that the results are both valid and reliable. I want you to be able to incorporate this monitoring into your routine without it becoming a pain.

1. Body Composition

Advanced

The most effective way to measure how our body is changing in response to training or nutrition is through measuring our body composition, which refers to the key tissues of the body, including body fat, muscle, bone mass and fluid.

For this book's purposes it is the body fat and muscle mass that interests us. They are usually reported as relative measures, such as **body fat percentage**. Elite endurance athletes can have body fat as low as 4 or 5 per cent for men, or 8 per cent for women; for the rest of us, a normal healthy range is between 10 and 20 per cent for men and 16 to 30 per cent for women. The lower end of each would be a more athletic physique.

There are many different techniques for measuring body composition, many of which sound suitably sci-fi, and all have their pros and cons. In sport we sometimes use DEXA (dual-energy X-ray absorptiometry) scanning, which hospitals use for measuring bone density (which we'll return to in Chapter 15, Ageing); this has the lowest margin of error. A big con of using this technique yourself is that it will likely involve paying for a private clinic – you're unlikely to find one of the machines at your local gym – and there is also the fact that it involves a small radiation dose, so wouldn't be suitable for everyone.

At the other end of the spectrum, skin fold measurements (anthropometry), which involves having your skin pinched with callipers (by an appropriately trained professional!), can be a reliable tool to assess meaningful change.

If you do have access to either of these methods, make sure you use the same method each time (even down to having the same person pinch your skin) so that you're comparing like with like, rather than using two different methods, which can increase the rate of error.

Basic

For the rest of us without access to body composition machines, it's time to dust off the tape measure and scales, and use our weight (as part of a *combined* approach, rather than on its own), waist size and our own eyes as our home body-composition assessment.

You should **weigh yourself** once a week, before breakfast. One of the biggest sources of demotivation I see is when clients weigh themselves each day – or worse, more than once a day. If you are a boxer or MMA fighter making weight leading into a fight, it can be justified to track small changes – but for everyone else, don't bother. It isn't going to monitor anything meaningful and it is more likely that it will induce stress as you constantly analyse any fluctuations in your weight.

Here's why your body weight can change so drastically from one day to another:

Hydration status: a 1 per cent loss in fluid equates to 1 kg in body weight (that's 0.75 kg for a 75 kg person) and can result from walking around for an afternoon and forgetting to drink fluids. As highlighted in Part I, when you start to exercise and sweat loss increases, it isn't uncommon for dehydration to increase above 2 per cent (1.5 kg for the 75 kg person). In women, the menstrual cycle can also affect fluid balance.

Carbohydrate storage: the fuel in the tank (stored as glycogen) actually binds with water too. With an average of 500 grams of glycogen (in liver and muscles), each gram binds with 3 grams of water so that, when fully fuelled, there is around 1.5 kg of water (depending on the size of the person). This explains some of the immediate losses made when someone starts a low-carb diet.

Gut feeling: waste products stored in the intestines before removal may add up to as much as 300–700 g to your weight, depending on your size. Low-fibre diets (called in this context low-residue diets) are used clinically (in hospitals before operations) and in sport (pre-competition) to reduce undigested residue in the gut.

Muscle growth: muscle tissue takes longer to build, so it takes longer to add 2 kg of muscle than to lose 2 kg of body fat. If you are training a lot, it is very common for lean mass to increase, especially early in a training programme – this often makes it feel like your weight is staying the same, even though body fat is reducing and muscle mass is increasing.

So instead of concentrating on your weight, think about your body composition. After all, it's the changes in both muscle mass and body fat that will help you perform better in training, look better in your clothes and improve your health long-term.

The next step is to **measure your waist.** As with weighing yourself, measuring your waist daily is pointless; you aren't going to see any changes overnight. Check once a month instead. This is how often we check body composition with our elite athletes and performers and it means there is time to see meaningful change. I'd recommend the same for when doing this at home – then put the tape measure away, along with the scales.

And, while this might seem like a simple piece of advice, not everyone measures the right part of their body. So, to be clear, to measure your waist:

1. Find the bottom of your ribs and the top of your hips
2. Wrap a tape measure round your waist, which is midway between these points
3. Breathe out naturally before taking the measurement

As a bit of a health warning, measuring your waist can give you some insight into whether you're storing visceral fat (the more dangerous, active fat). So regardless of your height or BMI, you should try to reduce your body fat (lose weight) if your waist is 94 cm (37 inches) or more for men, or 80 cm

(31.5 inches) or more for women[2], and you should see your GP if your waist is 102 cm (40 inches) or more for men, or 88 cm (34 inches) or more for women.

Use Your Eyes

This method isn't measurable in quite the same way, but it's still a reliable indicator that you're making progress. Every week ask yourself how your clothes are fitting. If your goal involves losing weight, then ask yourself if your trousers are feeling any looser round the waist, or if your shirts or tops feel any baggier. If you're trying to build muscle, are your trousers feeling any tighter around the mid-thigh? Shirts or tops pinching more at the chest, shoulders or upper arms? It might be subjective and lacking hard data, but often you'll feel these things as they start to happen – and that can be a more satisfying result than analysing any quantity of numbers. Use a trusted friend as a barometer – and not someone who discusses weight loss as a daily greeting. These guides, alongside weight and circumference, can form your home body composition assessment, and demonstrate you are moving in the right direction.

But as we've discussed throughout the book so far, changes to body composition are only part of the answer. What about how you feel each day?

2. Wellness Tracking

In Olympic sport, whether it's a judoka or a boxer making weight before the Olympic Games, a sprinter training to improve strength and power or a long-distance runner aiming to improve endurance, it is vital to know how athletes are responding to training (and life). This is also an essential piece of knowledge within football; with Arsenal Football Club and most international teams I have worked with, we regularly use wellness questions, in which we ask athletes to rate how they are feeling in terms of mood, energy levels, sleep quality and other key questions, on a sliding scale of 1 to 5, to gauge how the athlete is coping.

It's important to keep the burden low, and these questions have been refined from longer research-based questionnaires and some specific points relevant to the field added in. In each of our examples above, the athlete will see changes in their energy levels, mood, sleep, appetite and soreness as their training programme changes from week to week.

We would use these questions on players *every day* over a prolonged period of time to establish their baseline, and then look for any fluctuations that might demand an intervention. So, an easy one to spot would be a player who normally scores between 3 and 4 on his energy levels, but comes in the morning after a match and reports a 1. It's a trigger for a further conversation with that player to try to work out the underlying issue (how was his training the day before? Is his nutrition on plan with fuelling and recovery?).

This is exactly the same for you and your training and nutrition. As you start to tailor your nutrition to support your training and work schedule, you will start to see patterns emerging that you can act upon.

Now, I know what you might be thinking: in a world full of fantastic science and technology, with the bottomless resources of an organisation like a Premier League football club, why would we use something as soft as a questionnaire? And the answer is simply that it *is* scientific.

A recent systematic review (that is, one that is drawn from the peak of the evidence-based research that we will discuss more fully in Chapter 14, Supplementation)[3] concluded that 'subjective self-reported measures trump commonly used objective measures', which leaves little for me to add.

It's easy to forget that ultimately these are people we're dealing with, not machines with binary outcomes. Yes, we use the questionnaire as a first line and we can and do use data to further investigate and validate things, but no one knows better than you how your energy levels are.

So, when it comes to measuring your own progress with the Energy Plan, a wellness questionnaire might well be the best tool you have at your disposal, just as it is for the athletes I work with.

WELLNESS QUESTIONNAIRE

Rate each of the following from 1 to 5:

Energy
VERY TIRED   (1) (2) (3) (4) (5)   ENERGISED

Mood
HIGHLY IRRITABLE/   (1) (2) (3) (4) (5)   VERY POSITIVE
FEELING DOWN

Sleep
INSOMNIA   (1) (2) (3) (4) (5)   VERY RESTFUL

Muscle Soreness
VERY SORE   (1) (2) (3) (4) (5)   FEELING GREAT

Stress
HIGHLY STRESSED   (1) (2) (3) (4) (5)   VERY RELAXED

Adapted from McLean et al., 2010

When using this questionnaire to self-monitor, you need to establish your baseline. Use the first couple of weeks to assess where you normally sit on each of the scale, – doing it every day, at least for the first couple of weeks, will help you establish that – so that you have a guide to see when you fall out of your norm; for example you might be affected by extra stressors around travel, work schedule or life stress (which we cover in Part III), and when you might need to adapt your training or nutrition to help. It will also enable you to look back over time and see how your plan has positively impacted on these areas of your life and not just on how your body looks.

Basic

Once you have established your baseline, use the questionnaire **once a week** to see how you are feeling in the key areas. This is during your weekly **check-in**, as outlined below.

Advanced

Use the questionnaire daily, like the athletes I work with. Unlike checking your weight, checking in on how you're feeling isn't something that will inhibit your progress. But it's important to do it at the same time each day.

3. The Weekly Check-in

When I was working with Olympic athletes or footballers within a training venue every day, I was able to see them regularly to look at how their Energy Plan could be refined to meet their goals.

Football players are always busy with their various commitments, so sometimes we would have to be more creative on where we would catch up. In the gym, the restaurant, or I actually had one of my most productive consultations with Arsenal forward Alexis Sánchez in the sauna, discussing recovery nutrition following a match. This may sound extreme, but the point is that we needed to find time in his schedule for a check-in about his nutrition; you need to make time in your schedule to check in about yours.

One of the most common issues I faced outside of sport is that the people I worked with wouldn't spend any structured time reflecting on the previous week and planning for the week ahead. This meant that the good behaviours established in week 1 would be a distant memory when I next saw them, after a few busy weeks.

So I introduced the weekly 'check-in'. Make no mistake about it, this is the glue that binds your Energy Plan together.

Most Olympic athletes would keep a training diary (despite the wealth of technology at everyone's disposal, usually just a paper diary) in which they could add notes on their training, nutrition, how they were feeling, including wellness tracking scores. So this is what I recreate now with my clients.

As a bare minimum for the busy, you need to check in once a week. This means protecting time on your calendar (just like for your training), but you could do it on the train in to work or if you get a spare 15 minutes to sit and do it in your local café at the weekend (this seems a natural time to check in). You can use the chart on page 136 as a template:

Reflect on the past week

- How do you feel?
- How were your wellness scores over the last week?
- What worked well?
- What changes have you successfully made, such as consistent lunches or an extra training session?
- When did you deviate from the Plan and why? This can help you understand the things that affected your Energy Plan, maybe a work deadline or a friend's birthday.

Plan for the week ahead

What do you need to plan for the week ahead that will affect your Energy Plan? Work trip? Social engagements?

What do you need to execute your Energy Plan: are your exercise classes booked? When will you do the shopping? (see Chapter 10, From Plan to Plate).

How do you feel during training?

The Borg scale (also known as the RPE (rating of perceived exertion) scale) was created in the 1960s and has been used widely in sport ever since. The original was on a 6–20 point scale; a 'modified RPE scale' (1–10) has since been developed and is widely used in professional sport and health and performance clinics. The scale measures how hard you find a training session and can be a guide for a trainer on whether it is too easy or too difficult. In this case, we are using the scale to see how you feel (how much energy you have) during training after eating different meals (or performance plates).

Monitoring how you feel during training (just like you do with your wellness tracking outside of training) is a way to assess how your Energy Plan is working. For example, thinking about the same session each week (such as your exercise class on a Saturday morning), how do you feel after a fuelling meal compared to when you are training low? How hard does training feel when you have had a flat white an hour before the sessions compared to when you haven't?

You don't need to do this every session like an elite athlete. This is a tool for you to use during your training week. When you are making changes to your Energy Plan around your training it's another marker to help show yourself that you are moving in the right direction.

Rating	Description
0	Rest
1	Very, very easy
2	Easy
3	Moderate
4	Somewhat hard
5	Hard
6	
7	Very hard
8	
9	
10	Maximal

RPE scale, adapted from McGuigan and Foster

4. Hydration

In Part I we outlined that sweat rates per hour vary greatly between individuals, even during the same training sessions. Each one of our three sporting examples from the beginning of Chapter 4 will use this calculation to understand their individual sweat losses, and then personalise their drinking behaviours to minimise dehydration in both hot conditions (such as warm weather training camp) and cool (autumn and winter fixtures).

This should be the same for you, whether you are planning to run your first marathon or are just interested in how much you sweat during that intense HIIT session at the gym. Thirst is an important physiological cue to drink, but it doesn't provide the full picture on the level of dehydration.

Basic

Check your urine colour and volume, particularly pre- and post-workout. You can refer back to the section 'Water' at the end of Chapter 2 for more guidance.

Advanced

To calculate your own sweat loss, weigh yourself before exercising, making sure you are wearing minimal clothes. Then weigh yourself again after exercising, wearing the same amount of clothing. This gives you a broad measure of sweat loss, and you should then aim to replace 150 per cent of these losses. So, for example, if you have lost half a kilo during the course of your training, you should aim to drink three-quarters of a litre of fluid. This can be useful as a guide if you're worried you're drinking too little or too much around training.

If you wanted to take this further still, you could calculate your percentage dehydration. In sport, we do this because of the decline in performance significant sweat losses cause. Anything over 2 per cent would be cause for concern as it can reduce physical and cognitive function; that would be, for a 70 kg person, 1.4 litres of sweat loss.

But What About Measuring Calories In and Out?

You might wonder why, given that one of the principles of the Energy Plan involves achieving balance between your energy in and energy out, none of the monitoring involves counting your calories in and calories out.

Firstly, it's important to realise that much of the available data for calories in and out, the kind that is usually tracked by a wearable fitness device (or one on your phone), contains a lot of error – as much as 25 per cent when it comes to calculating energy out[4] – and the margin of error when recording intake can be as low as 2 per cent and as much as a whopping 59 per cent – a huge variability.[5]

Whether you're calculating your energy out with a fitness tracker or your energy in with a mobile tracker, it's important to realise that they're just estimates. They will contain error. They can be useful to give you a ballpark figure as to whether you're eating or moving more, but if you're doing too much maths around them, it might be an idea to call time on them.

I don't want you to be fastidiously photographing every morsel in your food diary, or frantically calculating calorie consumption on your calculator. Your energy will be better spent elsewhere. With all of our athletes, performers and businesspeople, results are created by getting the core principles right at each meal, each day, each week and monitoring the outcomes (with the wellness tracking and body composition described above).

As we've discussed before, looking at calories as your fuel budget for the day is a useful guide, but eating to an energy budget without focusing on the fuel composition can lead to insufficiency in some of the nutrients and micronutrients your body needs to function optimally. It's much more important to meet the targets of key fuels that the body requires, using the planners in Chapter 6. Remember the principle of the TTA model that we looked at on page 62: the **type** of food (its nutrition content, whether its role is fuelling or maintenance), the **timing** of the food (how close to training?) and the **amount** (portion size at each meal, and over the day). Because it's only through these that you can learn to achieve the sustained peaks and reduced troughs that we will tackle in the next chapter.

HOW DO YOU KNOW IF YOU ARE
'MOVING THE NEEDLE'?

The seven signs are:

1. **You have more energy**
   Feel more energetic throughout the day
   Less reliant on caffeine
2. **You sleep better**
   Get to sleep quicker and have better sleep quality
   Feel more refreshed upon waking
3. **You're in a better mood**
   Feeling happier and more positive
4. **You feel satisfied after eating**
   Not 'full' after eating
   Or hungry all of the time
5. **Clothes fit differently**
   Either a bit looser (fat loss)
   Or a bit tighter (muscle gain)

6. **Fitness has improved**
   You feel stronger during your training sessions
   (lower RPE)
   Improved fitness in aerobic sessions
   Lifting more weight in the gym
7. **The bottom line – more productive**
   Better in training
   Better at work
   Better at home

CHAPTER 9

On-Plan in the Workplace

S o far we have our goal, our performance plates, our planners and our monitoring strategies in place, but you might well be thinking: that's all well and good for a professional athlete or a performer who has their day mapped out and food prepared for them, but what about the demands the rest of us face in our lives? After all, we all face different challenges to our Energy Plans, whether that's a busy family or personal life, travel demands or work.

It would require a book far longer than this to deal explicitly with every individual challenge we face, but suffice to say the biggest inhibiter, beyond motivation or anything else, is time. Family and social commitments and the like all reduce the amount of time we're able to devote to our Energy Plan, but probably the biggest obstacle is your work.

This is where athletes and performers most definitely have the edge, because their Energy Plan essentially *is* their work.

Unless this is what you do for a living, your Energy Plan needs to incorporate your training, lifestyle *and* work.

In this chapter I'm going to loosely define working groups according to type (and this is by no means exhaustive – you may well find yourself straddling more than one type), and address some of the most common challenges (or, indeed, 'blind spots') you're likely to face as part of your Energy Plan, so that you can achieve the sustained peaks and reduced troughs you need.

Job Types and Daily Challenges

The four loosely categorised types of work are listed below:

Type 1: Flexible hours (entrepreneur, gig economy)

Type 2: Sedentary (office-based job, usually hinging around 9–5, Mon–Fri)

Type 3: Active job (forces, emergency services, building, some retail jobs)

Type 4: Shift work (hospital staff, night-shift workers, hospitality)

Of course, a self-employed designer or accountant may straddle types 1 and 2, while a night-shift-working doctor or nurse would certainly expect to find themselves included in types 3 and 4. And, while there's certainly some overlap, let's make the distinction between shift workers and flexible hours here: while the shift patterns of many shift workers certainly present a form of flexible work, they are usually done within a structure and as part of a broader need for staffing levels at various times of day and night. The flexible hours of type 1 are more to do with the *choice* an entrepreneur is able to make about when to work.

So, let's take a look at some of the common obstacles to your Energy Plan depending on your working type.

Lack of Activity

Who? Groups 1 and 2

Self-employment is on the rise. In the UK, the number of self-employed has risen from 3.3 million in 2001 (12 per cent of the workforce) to 4.8 million in 2017 (some 15.1 per cent of the workforce).[1] Advances in technology have played a huge part in this – we can be connected to work from anywhere. We've also seen shared workspaces popping up in major urban centres all over the country, as well as an increase in many organisations in home-working for at least part of the week.

All of which should mean a more empowering lifestyle, reducing stress levels and time spent commuting and, in theory, freeing up more time to spend training and preparing performance plates.

That's the theory – but the truth doesn't always work out that way.

A client of mine, Jessica, gets up on Monday morning. She has a quick shower and then breakfast as per her Energy Plan (overnight oats with a topping of chopped fruit, and a coffee), before settling down to her laptop at the kitchen table ready for a productive day at work.

Without the train ride in and the clamour and conversation of her buzzing office, Jessica finds herself flying through the morning, getting much more done than she would in the office. As she walks to the shop at the end of her street to pick up some fresh vegetables for lunch, she thinks to herself, I could get used to working like this all the time (after her Monday at home, she spends the rest of the week in the office).

*After having her performance plate for lunch – a piece of salmon with quinoa salad and Mediterranean vegetables, her **fuelling** meal for the day – she picks up where she left* *off, stopping mid-afternoon to make herself a coffee, and not long after she finds she's done for the day and has her whole evening in front of her. Brilliant. And then she checks the step count on her phone...*

It's only 30 per cent of what it would be when she's working in the office.

This is quite typical for the stay-at-home worker. With her journey to work reduced to simply walking down the stairs and into the kitchen, Jessica has little reason to leave the house. Even catching the train involves the walk to the station and in to work – incidental activity she probably takes for granted – and not only that, the walks on her breaks to get a coffee or to the photocopier or to see a colleague have gone.

Now, one day in isolation isn't going to make a huge difference, and Jessica has quite sensibly structured her Mondays as a **low day**, as she isn't doing any training, but what if that pattern were to repeat itself two or three days a week for the foreseeable future? What about someone whose entire working week is at home? Cumulatively, the days on which your activity levels are having more of an impact can derail the progress you make with your Energy Plan.

When working flexible hours, it's vital not to let your environment dictate your activity levels. I always ask my clients to mirror their usual work habits when they're at home. So, if Jessica takes closer to 10,000 steps a day when she's in the office, she should try to replicate that at home. And, if you work at home all week, it's definitely time to form new habits so that your incidental activity levels can increase.

The key is to **seek activity** throughout your working day, and see where the opportunities are to increase your activity levels.

Walking Breaks

Bank up your daily phone calls, then in the afternoon put on your hands-free and go for a walk around the park – or anywhere else, for that matter – while you make them all. (This is a principle that can be applied whether you're in a workplace or working at home.)

Hunting Lunch

Go for a walk to source your lunch. Or, if you insist on having it at home, ensure that you go for a walk round the block beforehand. (As above, you can apply this principle whether you're in a workplace or at home.)

The Commute

Can you squeeze a bit more activity out of your commute to work? Could you walk to the train station instead of getting a bus? Could you catch your bus or train a few stops further down the line? If you work at home, could you build a simulated 'commute' into the morning and afternoon, so that you go for a 15-minute walk to mark both the start and end of the day?

Of course, there will always be reasons not to do this: it's raining outside; I have too much to do to waste time on walking; I'll be late for dinner; I should make calls in the office in case I need to get information from my computer. But there are some very real reasons to do it too, and at these points it's worth examining just how important your goal is to you, going back through the reasons and motivation that you used when you set it in the first place. As we will see, your goal needs to become set in stone.

I have a client who regularly travels to India for work. He takes the car everywhere, to meetings and company visits every day, and can't get his activity levels up during the day. The solution: he walks on the treadmill in the evening when he gets back to his hotel, until he reaches his target. There's always a solution, even if it's not always fun, and with travel being such an important part of many jobs, it's important to look for options when you're on the road.

Underfuelling During the Day

Who? Groups 1, 2 and 3

I encourage the athletes I work with to fuel for the physical demands of their day, which means on most days increasing energy intake – in the form of lower-GI carbohydrates earlier in the day – for morning or afternoon training. Unless they are competing in the evening, dinner is generally lower in fuel.

But every time I take a look at the 'normal' working world, it feels like the message isn't getting through. Fuelling is more typically based on habit, and generally involves a light break-fast, a sandwich for lunch, no mid-afternoon snack and then being so hungry by dinner that anything put in front of us, usually the biggest meal of our day, will be wolfed down. It's back to front.

From a performance perspective this often means a lack of energy during the working day (and while training), meaning you are not operating at full capacity, and then consuming more energy in the evening, when you are less active. This can lead to choosing more energy-rich foods,[2] negatively affecting metabolism and weight management.[3] This daily structure also means we aren't getting the best **maintenance**. Insufficient protein intake at breakfast is common, even among elite athletes.[4] This means that both energy levels and muscle repair take a hit during the day. It's bad news all round.

Flip your current pattern: start by reducing the carbohydrate at dinner. This will be key to creating an appetite in the morning. Then front-load your daily energy intake: aim at breakfast and lunch being the biggest meals of the day and evaluate how you feel using the monitoring questionnaire on page 136.

Be aware that these changes won't happen overnight. Give your body three to four weeks to adapt to this new pattern.

Keeping Your Health Appointments

Who? Groups 1, 2 and 3

Timing, which we discussed in Part I, is the crucial bridge between training and nutrition: eat at the right time to keep your blood glucose stable and feel energised for your session. If you eat at the wrong time you feel either too full and uncomfortable because you've eaten too close to training or, if it's been too long, weak and hungry.

Let's meet another client of mine:

Ryan is a business development manager for a new start-up. His day involves generating new business both on the phone and at meetings across the city. This generally means lots of time spent on the train.

Ryan is motivated but not in control. He goes to the gym every other day, but mealtimes are extremely inconsistent – because of his meetings, lunch gets pushed back or even missed. And come mid-afternoon he often succumbs to the snacks made available for the team in the office – the usual mixture of chocolate, biscuits and cake.

He generally intends to fit his workout in after work, but by this time the day has often run away with him and he either has another meeting or is too tired to train.

I'm sure this is a common issue for many readers. And it quite simply comes down to treating your health as you currently do your business appointments – prioritising it.

Lunch and your training have to become appointments in your week. Depending on your goals and health status, they may well be the most important, so they need to be treated as such.

Put lunch and training sessions in your smart phone's calendar alongside your other meetings (see opposite). They can also be in your diary or on the calendar on the wall at home, but they need writing in proverbial stone. Do it during your weekly 'check-in', which means prioritising them and scheduling other meetings around them. Treat them as immovable appointments and make this clear to colleagues and even loved ones.

Eating Late

Who? Group 4 (and potentially all)

For some industries, the business end of the day is during the evening. The hospitality industry, for example, comes alive at night, and plenty of people go to work just as the 9–5ers are heading home. In this instance, as with anything in your Energy Plan, the question comes back to what you are fuelling for. In this case it's for the evening, so, as it would be

Keeping your health appointments

for a footballer playing in an evening kick-off or a West End performer, the meal eaten pre-performance needs to provide the right amount of energy for the work at hand: the **fuelling** performance plate should be used.

Let's say you work in hospitality, and your shift finishes at midnight and you're back home by 1am. Do you eat then? Is it *too late* to eat? In fact, as we discussed back in Chapter 5, having a protein-based snack helps maintain and repair muscles, so my advice would be either to eat a protein-based light meal (a **maintenance** performance plate) or snack, as extra fuel (carbohydrate) won't be required ahead of when you next eat in the morning. If you have had a particularly busy shift, and have been on your feet all night, then some carbo-hydrates could be added to make it more of a fuelling option and replenish your glycogen stores.

With a football club playing an evening game we have two post-match options available for the players. We have the fuelling option, with carbohydrate, for those who have played in the match, and a protein-rich maintenance option for the unused substitutes. So at the end of your shift, think of your activity in these terms: did you play or were you sitting on the proverbial bench for much of the night? Fuel yourself accordingly.

Of course, the infrastructure needs to be in place what-ever kind of meal you're having, if it's at 1am. It requires organisation and planning so that you have the right types of foods to prepare either a maintenance or fuelling perfor-mance plate. And don't believe any of the myths surrounding eating late – if you are an active person, skipping a late meal when you need it is one of the worst things you can do when it comes to your Energy Plan. In Part I we established the importance of recovery and muscle protein synthesis, and this process continues overnight while you sleep. If you've had an active evening, whether in the gym, on stage or running the floor in a restaurant, going to sleep without having eaten will leave your muscles in negative equity. If you're worried about eating too close to bedtime, leaving an hour after you've eaten will be adequate.

Being a Creature of Habit

Groups: All

We are creatures of habit. The average Briton has just nine recipes in their cooking repertoire.[5]

Many people have well-worn habits: during the working week it's the same breakfast on the go, sandwich from the same shop and a few dinners; then at the weekend it's all about eating out or takeaways.

But variety is vital. It's vital for ensuring we have the full range of micronutrients in our diets – in the Immunity chapter we will talk about the importance of eating a variety of these protection foods. We have already seen that different meats, grains, pulses, fruits and vegetables all provide different vitamins and minerals, crucial for the repair of our tissues. And although the focus of the Energy Plan is often on fuel to meet our energy demands, the importance of variety in its composition shouldn't be forgotten.

Ensure that you eat different types of proteins, pulses, grains, fruits and vegetables to increase your dietary variety. I also encourage anyone I work with to try out a range of new recipes to broaden their repertoire.

Constant Snacking

Groups: All

I see a lot of people fall into the habit of constant snacking, particularly at work, and it is often driven by the perception that they're too busy to eat a meal. The outcome of this is typically erratic energy levels and mood created by the constant blood-glucose rollercoaster, and subsequent underperforming at work. And their reduced appetite when mealtimes come means they don't feel like eating a performance plate full of the right fuel for their needs.

Start by getting rid of the snacks and just eating three meals, with the relevant portions and fuels for your needs (I recommend starting with a medium day without snacks). Track your wellness measures. The drive to eat a full meal is now there, and your energy levels will become more consistent. Also keep a note of how your appetite is changing during your weekly check-in. Overall you should become more productive at work and at home.

BLOOD GLUCOSE AND MOOD

Eating meals at irregular times (or skipping them) or eating meals based on high-GI carbohydrates or sugary snacks can cause big swings in blood glucose (the sharp spike and fall that we discussed with high-GI foods on page 25), which as well as affecting energy levels also affects mood. The elevation in blood glucose from high-GI diets with refined carbohydrates is linked to increased anxiety and symptoms of depression.[6] Rates of depression are also higher in those with diabetes.[7]

Focusing on meals with lower-GI carbohydrates and eating at regular times as part of a 24-hour fuelling strategy will reduce spikes in blood glucose; see the diagram on the next page.

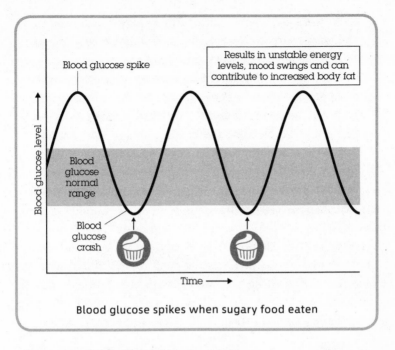

Blood glucose spikes when sugary food eaten

The Law of Diminishing Returns

Groups: All

This describes the point at which the level of benefit from something in your Energy Plan becomes less than the amount of time, energy or money you put into it.

Ella is a digital analyst for a marketing company. She spends much of her day poring over data and analysing numbers to find the story behind them, and in the past she's also brought this almost obsessive level of detail to various fad diets she's tried, leaving her with severe peaks and troughs and doing way too much maths around her calorie consumption.

Ella wants to change and do something more sustainable (and less obsessive) with her nutrition. Over three months she makes great progress with her Energy Plan – lunch and dinner have both been switched to include lower-GI carbs and higher protein intake and timing is now like clockwork with meals and snacks. Her energy levels are more stable. She recovers more effectively from her training now and, as an unexpected bonus, she's lost 2 kg of body fat.

In fact, for Ella it all feels a bit too easy. Where's the jeopardy? She is used to feeling hungry and tired and counting each calorie – and without the constant tracking she feels that she could be doing more. She wants to start tracking calories and macronutrients, weighing every gram of her food in an attempt to be even more precise and elicit even bigger gains. Before long, she can feel herself slipping back into old obsessive habits...

As we discussed earlier in the book, the data around calorie consumption and expenditure is open to substantial margins of error; trying to be too precise with this probably flawed data will just lead to you putting in more and more effort for fewer gains. So don't waste your time (and energy) on it.

Try not to sweat the small stuff. Ella's focus instead should be to track her wellness, fitness levels and even body composition – the outcomes from her Energy Plan. Through doing this she can start to get a better understanding of her training and nutrition, how she is feeling every day, her mood and sleep quality. She can start to see patterns here, such as where she is becoming more consistent in her response to training and when she has sustained levels of energy.

The Dose makes the Poison

Groups: All

Josh is super-engaged with his health and well-being; I can tell from the microscopic detail he has chosen to share on his pre-assessment forms. A snapshot of one of his days includes the following:

- *Juices: morning cold-pressed juice containing apple, spinach, pear and banana*
- *Mid-morning snacks: energy balls*
- *Post-training shake with almond milk (and sometimes a dollop of nut butter)*
- *Handful of nuts mid-afternoon*
- *Plenty of coconut oil when cooking at dinner time*

It's almost like a who's who of health trends. But there's one issue. Despite training every other day and following a very active lifestyle, he seems to be gaining weight.

Unfortunately, all these 'healthy' bolts-ons add up. Should he be including some of these things? Yes, absolutely. Does every meal and snack need to include them? Absolutely not.

Many people would say that diet looked healthy – but that's the issue. Out of context, all these bolt-ons are not necessarily 'healthy' and the word is not meaningful. **'Healthy' foods can still make you fat – your metabolism won't discriminate.**

One of the key principles for Josh was to increase his amount of protein to support his training – and by reducing fat intake (even though he was eating mainly healthy fats) and sugars (and overall calorie intake), his body fat started to reduce.

The broader question to ask before you integrate the latest hot health food into your diet is: 'How does this help with my Energy Plan?' If you can't answer this, then what is it doing there?

CHAPTER 10

From Plan to Plate

'**I**'ll give it a seven,' I said.

This was my first food tasting with Raymond Blanc, and you could have heard a pin drop as I delivered my score out of ten to him and his chefs. We were collaborating on a year-long project to upgrade the match-day recovery food at a football club. Our aim was to see our nutrition targets for the players translated into the best fresh meals, with responsibly sourced, seasonal ingredients. The project delivered the perfect marriage of performance and great-tasting food.

However, at the time all I could think was, *I can't believe he made me go first.*

I have been fortunate enough to collaborate with some excellent chefs through my work in sport, but it's Raymond Blanc's philosophy and attention to detail that I try to take with me when we reach the stage of the Energy Plan that sees us bridge the gap from **plan to plate**.

The Performance Chef

I've always hated the concept of bland food in Tupperware. It is possible to enjoy good food and still achieve your goals; and in fact, if you are working hard, it seems to me to be even more important that food is a pleasure.

As you'll be well aware of by now, the athletes I work with aren't on diets. They are instead given the tools to help them meet their goals, and this includes cooking skills. At Arsenal we used to run cooking classes with the club's head chef Christian Sandhagen for all the players, particularly those new to the club. We would work through everything from online shopping to food hygiene, to chopping and knife skills – *very* important for multimillion-pound players to get right – and would teach them to prepare meals according to their goals and activity levels, just as you learned in the Performance Plates chapter. The most important aspects, however, were to make it fun and as straightforward as possible and to engage the players.

Ten years ago in sport it was a different story. Nutrition didn't carry the weight then that it does today, so there would only be one chef cooking for the players and cooking 'healthily' was understood to be keeping the food low-fat.

Today I can be working regularly with several chefs – at the training ground and stadium and the exec-travel chef – and this doesn't even include all the personal chefs working with individual players, which is a growing area of the industry and really highlights the importance of food in sport.

A personal chef to a football player is cooking for a finely tuned athlete, not a fine-dining restaurant, so they need to be able to provide performance plates in the form of both meals and snacks that fit the right nutrient profile and quantities. And the best chefs are great problem-solvers, and are able to field these sorts of questions from me:

'Can we get more protein in that snack to reach 20 g for muscle recovery?'

'Can we increase the levels of antioxidants in the salads and drinks after the match, to support repair from muscle damage?'

'Can we create a jelly shot high in vitamin C and gelatin to increase collagen synthesis and support tendon health?'

I even have conversations with suppliers about the dietary nitrate content of their vegetables and juices and how the soil is monitored. *The details matter.*

I say all this not just to highlight the impressive resources available to us in professional sport, but to highlight the importance of good-quality, responsibly sourced and tasty food to the Energy Plan, alongside its core role as a fuel.

A lot of time is put into developing the kitchen skills of young athletes I work with, so that they can prepare the meals they require to recover when they get home. Some of the world's most confident and proficient athletes start off having no idea how to put meals together – but then, unless they're shown the ropes, many people don't. And like any new skill, confidence in the kitchen takes times and effort to build.

As a starting point I'd recommend a selection of core, versatile recipes.

As mentioned in Chapter 9, the average person has nine recipes in their repertoire. Some people have many more, of course, but the truth for many of us is that, while we might like the idea of cooking, we don't always have the time to scroll through websites and then go to multiple shops to buy the ingredients.

It's my job to assess what the limiting factors are to a performer being able to execute their nutrition strategy, and with the Energy Plan the most common limiter is, of course, time. So in this chapter I'm going to show you some strategies to shop smarter and set up your kitchen more efficiently, and some of the key factors I used to build my recipe collection, so that you can become the **performance chef** in your Energy Plan.

Your starting point is your weekly check-in. No matter how busy your week, this slot is essential as an anchor to your week (refer back to the chapter on monitoring for a reminder of the details). As well as reflecting on the previous week, this is when you plan the week ahead – it's the first step in making the journey **from plan to plate.**

CULTURAL AND DIETARY DIFFERENCES

Multiculturalism helps shape food trends in countries around the world. In cities like London there is a thriving restaurant and food scene, with restaurant-goers having a choice of cuisines from countless different countries and regions: from Afghan to Peruvian and just about everything in between.

We see this in professional sport too, especially football. At Arsenal Football Club we had 16 different nationalities in our first-team dressing room.

I have also worked with the BBC's GoodFood. com to create a Marathon Hub for recipes and nutrition information for the London Marathon; another sporting event showcasing a wide range of nationalities and food requirements and preferences.

In both these cases, the solutions and meals we provided needed to appeal to a range of cultures and dietary needs. The guiding principles for all the plans were exactly the same – we just used different foods for the solution.

In elite sport the challenge is to ensure that athletes get all the relevant fuel necessary in accordance with their dietary or cultural preferences. It's certainly easier to provide fuelling solutions for someone who eats meat and has no restrictions, but there's always a way to provide an answer for athletes who have other preferences.

And with the Energy Plan, your food solutions too will be tailored to your individual cultural needs, as well as your likes and dislikes and dietary needs such as vegetarianism or food allergies.

Quick and Easy: The Midweek Recipe Solution

I have always worked with people who are time-poor, which is why one of my biggest frustrations earlier in my career was recipes with huge lists of difficult-to-source ingredients; it just reduces the likelihood of adding new recipes to a limited repertoire.

So, when working with a chef to create new recipes, this is the brief I set out, the same that I used with the England football team's chef Omar Meziane. This brief produced the recipes currently available on my website:[2]

1. **Up to four versions.** Each recipe can be made and enjoyed in up to four ways - an easy way to add to your repertoire.
2. **Minimal ingredients.** What is essential to the dish? This is what we always work back to. Unless it's essential it's out.
3. **Quick and easy to prepare.** Ideally under 20 minutes and involving just a few steps.

I would encourage you to use some of these pointers yourself when building up your recipe collection to use at home. A professional chef may not be at your disposal in person, but plenty of recipes by professional chefs most certainly are, on the internet and in many cookbooks. Resources like the BBC GoodFood.com recipe hub are a good starting point. Look for recipes with **minimal ingredients**, those that are **quick and easy to prepare** – almost all recipes have a guideline as to the cooking and preparation time – and, most importantly, using what you've learned so far with your Energy Plan, evaluate whether this meal is composed of the right types of fuels for you.

Look at whether the recipe can be made in more than one way. If it's a chicken dish, could you substitute that for fish?

For vegan dishes, could you try a different grain or pulse as the protein source? As we've discussed throughout this book, the starting point for every meal is the protein, so look at meals built around things like tofu, eggs, pulses, fish and meat, depending on your dietary requirements.

Cooking can be fun and the rewards for your Energy Plan can be huge, so trying new dishes and adding to your recipe collection can help you learn, grow and stay engaged with food. In my experience, I've found that pasta, rice, omelette or stir fry are things that most of us can cook, but they can become a default option when the cupboards are bare, the family are hungry and time is of the essence. So let's look at the habits we can build with our Energy Plan to ensure our default options, with the narrow variety of fuel that involves, don't become too regular an occurrence.

Performance Shopping

Some people love a visit to the shops, browsing the aisles and trying new foods for new recipes, while others hate it, making smash and grab raids on the local 'metro' supermarket most nights of the week, often while on the phone and very hungry. That last one is a particularly easy trap to fall into.

There are now more tools than ever at our disposal to make food shopping easier. Online supermarket shopping takes a lot of the hassle out of it and means your shopping can be done from your desk without eating into your schedule, and there are a whole host of companies that will send you boxes of ingredients and recipes, with all nutrition information listed. These can be enormously helpful, provided the recipes fit your goals and fulfil the criteria of performance plates.

But whatever your approach to shopping, here is the Energy's Plan's three-step approach to elevate getting your groceries to **performance shopping:**

Step 1. Kitchen Essentials

These are multipurpose ingredients that you'll use all the time. Buy them in one big shop (stock up on utensils and storage containers here too):

- Grains and seeds
- Pulses and tinned goods
- Spices
- Oils and fats

Step 2. Sunday – The Primer

Spending an hour on this on a Sunday* will stop the endless supermarket visits during the week and give you ownership of the week ahead:

- Fruit and vegetables
- Dairy
- Wraps and bread
- Meats and fish
- Herbs

Step 3. Wednesday – The Refresher

This is the quick Wednesday-night visit to top up the fresh items – just 20 minutes in a convenient supermarket or other food store:

- Dairy
- Fruit and vegetables
- Specific ingredients for new weekend recipes

* This assumes you're a Mon–Fri worker, but obviously choose any convenient day for you.

PERFORMANCE SHOPPING HOME ESSENTIALS

- Water bottle (with a fruit infuser, if you prefer)
- Protein shaker
- Coffee capsule machine (to understand your caffeine dose and timing, and keep it consistent)
- Storage boxes (for lunch and snacks at work)
- Kitchen essentials: sharp knives, chopping boards, non-stick pans, griddle, saucepans, baking trays
- Heavy-set glasses and cutlery (see Chapter 7, Winning Behaviours)

Meal Planning: One Step Ahead

With your performance shopping you're in the driving seat to deliver everything your engine needs for the week, and doing some preparation on a Sunday, such as ensuring you have all the ingredients, or even preparing part of your meals for Monday and Tuesday, can only make the start to the week easier. So here you might want to break out the food containers, which is fine as long as you remember it's what's inside that counts: make sure they are filled with food that is new, interesting, exciting – or reassuringly tasty and familiar – whatever it takes to make that meal an enjoyable experience for you.

You'll already have your week mapped out in terms of your social and work engagements, your training plan and the types of days you need on each – high, low or medium days – and now you can spend a bit of time looking at which meals you could get a start on preparing to keep in the fridge on a Monday and Tuesday night. As a first step, just having the foods at home will make it easier to stay on plan. Your Wednesday 'top-up' shop will ensure you have fresh vegetables, fruits and other ingredients so you can try out some new recipes with family or friends. With just a little preparation at this stage you can avoid having to eat into your pinched after-work time and instead, when you've completed your training, step into the kitchen to meals you've already made a start on. Any leftovers can also work well as lunch the next day, so it's well worth investing some extra time here.

And if you prefer to cook as you go, that's fine too. Just having the ingredients ready to go in your kitchen will make a big difference, especially if you normally find yourself having to spend time visiting a food shop every night after work.

SHOULD I BE EATING ORGANIC?

The Soil Association's 2018 Organic Market Report revealed another 6 per cent rise in consumption of organic food and drink, the sixth consecutive year of growth in the market. Total sales were £2.2 billion.[3] It's a similar tale in the US, where the Organic Trade Association reported a 6.4 per cent rise to give total sales of $49.4 billion (£37.5 billion) in 2017.[4] Organic is big business, clearly; but the debate about its benefits has rumbled on for years, with lots of opinions and, unsurprisingly, lots of conflicts of interest, just as with supplementation (see Chapter 14).

With the Energy Plan we always look at the quality of the evidence first, and one systematic review (review of studies that meet a certain standard) highlights the number of variables that can affect the nutrient content of both crops and livestock before they reach the plate. It concluded that there are no significant differences between organic and conventional produce in terms of nutrient content.[5]

However, a wider discussion point is the environmental impact and how you source your food. Wherever possible - depending on budget and other factors like convenience, of course - sourcing food locally and seasonally, prioritising sustainable standards, reducing packaging and waste are all steps we can take.

I often warn my clients about the 'halo effect' of language such as organic, healthy and gluten-free. These foods still need to be tailored into a plan. We talked earlier in the book about how you can overdo it with too many healthy foods eaten indiscriminately; and bear in mind the mantra: *if it's not helping you reach your goal, what is its role?*

PART III

Sustainable Energy

CHAPTER 11

Recharging

S leep is one aspect of performance where the difference in attitude between the sporting world and the 'real' world couldn't be more different. In professional sport sleep is viewed as a necessary, essential aspect of recovery and repair, revitalising the body and mind so that a performer can recover from competition and switch off away from the glare of the theatre in which they play. Sports stars such as NBA star LeBron James, Olympic sprinter Usain Bolt and tennis champion Serena Williams speak openly and boldly about their need for sleep. At *least* eight hours a night, and often more. Sport harnesses the competitive advantages of getting sleep right, with sports science teams always looking at sleep to improve the performance of their athletes. As with nutrition, if it has an impact on the bottom line – performance – then we extract every benefit we can from it.

Compare this with the traditional outlook in the world of 'normal' work. It's long been considered a badge of honour to declare how little sleep one needs; there are plenty of examples at the higher end of the achievement scale to back this up. People like Indra Nooyi, former CEO of PepsiCo, and former US President Barack Obama, who each slept for only around five hours a night. Fashion designer and film-maker Tom Ford gets even less, often as little as four and a half hours, while former British Prime Minister Margaret Thatcher famously had a paltry four hours a night.

Which end of the spectrum do you find yourself at? Long hours at work, training either before or after, an active personal life... and that's before considering those of us who have young children – it's a wonder any of us find time for sleep at all. And, given the options of going to a new restaurant with friends or having an early night to prioritise sleep, few of us are likely to choose the latter. Professor Matthew Walker, Director of the Center for Human Sleep Science at the University of California, Berkeley, describes how people confess to him that they 'need eight or nine hours a night' as if it's something to be ashamed of.

Things are changing, however, with people like Walker and Arianna Huffington – founder of the *Huffington Post* and self-confessed seven-hours-a-night sleeper – leading a charge to reclaim the night and make time for sleep. Walker points out the perils if we don't: 'No aspect of our biology is left unscathed by sleep deprivation.'[1] This means things like type 2 diabetes, cancer, heart disease and, in the long term, conditions like Alzheimer's and obesity.

In fact, sleep has become a buzz lifestyle topic, with people looking to address not only their sleep quantity, but its quality as well, in order to get that pitch-perfect harmony between stress and rest. For what it's worth, the National Sleep Foundation in the US recommends a minimum of seven hours' sleep a night[2], but each of us is different, and you might need more. Making room for at least this amount of time to sleep would be a good start to address the quantity if you currently get less.

From a nutrition perspective, I am constantly looking at the evolving relationship between nutrition, sleep and rest. Sleep is an area about which we still know relatively little, but some of the biggest brains at universities like Stanford, Munich and Oxford are working on changing that and there is a burgeoning amount of research in the field, which brings regular new developments.

As with many aspects of nutrition, the desire to see results – striving for better sleep quality and quantity – leads to many people investing their time and energy in the wrong areas, including false friends such as supplements, including herbal products, which are subtly (and not so subtly) pushed as a 'magic bullet' for better sleep. When it comes to your Energy Plan, it is better to focus on getting your nutrition and activity needs in line first before looking into the 'marginal gains' to be made elsewhere.

Remember the four Rs of recovery from Chapter 3: **Refuel, Repair, Rehydrate** and **Rest**. Having dealt with the first three, we should now look to the last of these to see how we can recharge our body in the right proportions, just as we fuel ourselves using the correct composition and dose, to give us enough energy to thrive while we tackle our goals.

Sleeping Like an Athlete

Sleep is a cornerstone of any athlete's recovery process. It's vital to the regeneration of cognitive and physiological function, and not getting enough can prove fatal to a performance: things like shot accuracy, decision-making skills, memory, pain perception, strength and endurance are all affected.[3] And sleep doesn't only provide for muscle and joint recovery. It plays an essential role in memory consolidation and learning; the development of new skills that has begun on the training ground is still going on at night while an athlete sleeps.

Every team and sports institute I have worked with has put an emphasis on nailing the basics around **sleep hygiene** (the term sleep scientists use for good sleeping practices). Sleep is an area touched by all the various sports science and medical teams: nutrition (what to eat, when and what to avoid), coaching (overall training volume and scheduling of train-ing to allow for recovery), psychologist (specific strategies to improve sleep) and finally the doctor to provide, in extreme cases, medication.

Below are the most common sleep hygiene principles and techniques drawn from high-performance sport. Some of them might sound like common sense, but it's amazing how many of us 'know' about them and yet still don't follow them. It's vital to get these things right to give ourselves the best opportunity for a good night's sleep.

SLEEP HYGIENE PRINCIPLES

1. Have a sleep schedule and, if your lifestyle allows, try to go to bed around the same time every night and get up at a similar time each day (easier said than done if you have young children, of course).

2. Make sure the room you sleep in is dark and cool. Our body temperature drops a degree or two when we go to sleep at night, and a room that's too warm could interfere with that and make it more difficult to get to sleep. Around 18°C is good, but it shouldn't feel cold.

3. Keep gadgets such as smartphones out of the bedroom – the blue light they emit can interfere with melatonin production (the sleep hormone) and inhibit your sleep. Engaging with apps and social media stimulates your brain and will keep you awake.

4. Don't drink caffeine too close to bedtime. Depending on individual tolerance this might mean no later than late afternoon.

5. Avoid bright lights in the evening – use warmer bulbs and lamps and lower lighting levels instead.

> 6. Find a way to relax immediately before bedtime, away from screens. Meditation or breathing exercises are popular options.
> 7. Leave *at least* 30 minutes after eating before you go to bed. This includes eating a snack after an evening workout, to allow digestion to begin.

Sleep is vitally important and forms a core part of the wellness monitoring strategies we use in sport – we track our athletes' subjective sleep quality, energy levels and mood status. When sleep 'goes into the red' on consecutive nights, it manifests itself in a very clear way – it produces a significant reduction in performance.

And it's exactly the same when applied to any of us. In your own wellness questionnaire on page 136, I ask the same kind of questions of you as I would of a professional athlete, because, if you don't get enough sleep, your decision-making skills, memory, accuracy and endurance will all be affected. This goes for your life outside of exercise and training too – any new skills you might be developing in your job, on a training course or in an out-of-work environment, such as learning a new language, are consolidated and stored in your memory at night when you sleep, just like an athlete consolidating new skills learned on the training ground.[4]

Disturbing Sleep

While many of us, as 10pm approaches on a Wednesday evening, might be comfortably settled down and thinking about bedtime, a Champions League footballer will just be returning to the dressing room after the match. He then needs to have his treatment from the medical team, shower and have his recovery food, meaning that he won't leave the ground until eleven and possibly won't be home until midnight... if it's a home game. And even then he'll be so wired from playing that sleep won't come any time soon.

One study of footballers showed that players in night matches got almost three hours' less sleep than if they had played a day match, and their subjective feeling of being rested was significantly lower.[5] For you in the 'real' world, working late or on varying shifts, or exercising in the evening, may well be affecting you in the same way.

Don't jump straight into bed when you get home after work or exercise expecting to sleep. You're likely to be stimulated and need to wind down just like an athlete or a performer. Your own way of doing this will be personal to you, obviously, but it is important to spend some time relaxing and unwinding, as well as giving yourself a chance to digest your food, before bed.

It's not just sleep *after* your performance that can be affected, either. There is good evidence from both the performing arts and sport that sleep is disturbed during the training and rehearsals leading to a big competition or production.[6] I'm sure many of us can empathise. The equivalent for us might be experiencing disrupted sleep on nights in the lead-up to a big event like having to deliver a best man or woman's speech at a wedding, an important job interview or the major endurance event we're nervous about competing in at the weekend.

In many high-performance sports facilities across the US and Europe, sleeping accommodation is provided for the athletes. Some football clubs, like Real Madrid, actually have apartments for their players to relax and sleep in. Other places offer things like sleeping pods and 'recharge' rooms, an approach that has been adopted in many business environments, which are starting to react to the impact of burnt-out employees on their business.

So sleep is clearly thought of as vitally important in the world of professional sport and in more progressive workplaces, as well as to all of us in general well-being terms. It is also very relevant to your Energy Plan in its potential to seriously derail your goal, particularly if it involves the most common goal for many of us: fat (weight) loss.

Sleep and Your Weight

The hunger hormones leptin and ghrelin are two of the main players in a multifaceted and complex production your body knows as appetite. Ghrelin is released by the stomach to signal to the brain when you are hungry, while leptin is released by fat cells to tell the brain that you are full. Research has shown that a lack of sleep raises ghrelin while reducing leptin – in other words, it increases your appetite.[7]

Not only that, but a lack of sleep plays havoc with your self-control. Not getting enough rest affects the frontal lobe, the part of the brain that controls decision-making; furthermore, when our sleep has been restricted, food triggers different reward centres in the brain. You're drawn towards more high calorie, sugary and fatty foods. Ever walked past a bakery or coffee shop on the way to work after a bad night's sleep and found the smell of sugary baked goods almost irresistible?

So it's hardly surprising to learn that being sleep-deprived leads to greater calorie intake – which can really mess with your Energy Plan. And the effects of sleep deprivation don't stop there. Lack of sleep can slow down your metabolism and reduce your energy expenditure,[8] which is believed to be your body's mechanism to conserve energy, and it also knocks your body's ability to manage glucose and consequently insulin levels.

If you aren't making enough time for sleep, or you aren't putting measures in place to get a good quality of sleep, you are fighting your body. This is a huge handicap to attempt the Energy Plan with, especially if your goal involves reducing your body fat. So what can we do to improve our sleep, aside from making more time for it? Let's have a look at what nutrition has to offer first.

In Chapter 2 we talked about warning signs like 'RED-S' – caused by either training too hard, dieting or a combination of the two – which knock your physiology sideways. Sleep quality is one of the things that can take a hit, and it is yet another reason for planning when to reduce body fat. I work with my clients to assess clear times throughout the year when we will target fat loss. This is generally at times when stressors in work, travel and other aspects of life are low.

What Can I Eat to Help Me Sleep?

As I work with a number of clients with atypical working schedules and irregular sleeping patterns, I'm often asked, 'What can I eat to help me sleep?' I think the expectation is often that we can simply introduce something into our diet to provide that magical eight hours – but in practical terms it's often just as much about taking things out of our diet. Let's take a look at some of the things we can do with nutrition to help our rest – and the hindrances we can cut out.[9]

It is of course vital to bear in mind that, while there are some promising results with some of the nutritional aids we're going to look at, the bank of evidence to draw upon isn't huge as research in this area is still in its early stages. So don't get your hopes up too much – nothing here is going to magically fix your sleep, but some of these things might well help, in combination with some of the sleep hygiene principles on page 176.

The Promotors

Generally speaking, diets high in **protein** may result in improved sleep quality and diets high in **carbohydrate** may result in shorter sleep latencies – that's sleep-science speak for the time it takes to fall asleep.

There are more specific substances that may promote good sleep too. One of the most interesting developments of recent times is in foods containing **tryptophan**. Your body's ability to synthesise serotonin – the neurotransmitter that plays a leading role in dictating mood – is dependent on the availability of tryptophan, which is an amino acid, part of your maintenance (protein) intake in your food. Tryptophan and serotonin also play their part in producing **melatonin**, the sleep hormone, which is released by the brain when it gets dark.

Doses of 1 g of tryptophan – the equivalent of 300 g of turkey or 200 g of pumpkin seeds – help to improve sleep latency and subjective sleep quality.[10] The Energy Plan's food-first approach means that you should aim to get as much of your tryptophan as possible from complete protein foods, in other words those that contain all essential amino acids; eat foods such as meat, poultry, eggs, soybeans, nuts and seeds in each meal or snack (especially at dinner and as an evening snack, if required), before considering supplementation.

Montmorency tart cherries are part of a group of foods containing **melatonin**. They have recently garnered a lot of attention for their potential to play a major role in recovery from heavy training thanks to their antioxidant and anti-inflammatory properties. There is some initial evidence to show that the juice raises melatonin levels and is beneficial to sleep quality and duration.[11] One dose is 30 ml of concentrated juice, also produced in capsule form, which helps to reduce sugar intake. You should start with one dose an hour before bedtime.

The Inhibitors

When **total calorie intake is decreased,** sleep quality may be disturbed. So an energy deficit to lose body fat should be managed carefully.

Diets **high in fat** may negatively influence total sleep time. Follow the principles in the **Performance Plates** (see page 79) to gauge the fat content of your food.

Caffeine, which we covered in Chapter 5, can interfere with sleep depending on your tolerance to it and the time of day it's consumed. Caffeine has a half-life of three to five hours, so that 6pm coffee could still be having an impact when you go to bed.

Alcohol, although it might help you to fall asleep faster, can reduce restorative REM sleep – which is vital to memory consolidation – and also impacts sleep patterns and quality. A recent sleep study carried out in Finland demonstrated that as little as one drink was shown to impair sleep quality, and that the more you drink, the greater this negative impact is.[12]

I'm not for a second suggesting that you must cut alcohol completely from your Energy Plan. As we discussed in Chapter 5, a well-earned drink is part of living well and – within government guidelines – can be a safe part of an active social life. Refer back to 'Rocket Fuel: Alcohol' on page 99 for a reminder of how to plan for nights out and incorporate sensible alcohol consumption into your Energy Plan.

Muscle Maintenance While You Sleep

Considering nutrition's role in sleep doesn't just mean eating food that can help with sleep itself; it can also mean consuming fuel that can go to work on other aspects of your body's needs *while* you sleep. It's the equivalent of putting the car in the garage for maintenance overnight. So let's take a look at our diet's main source of maintenance: protein.

Protein is a must-have for your evening-meal **performance plate**. And you may be having another protein dose later in the evening, if you have a heavy training session and have your dinner early. In the protein section in Chapter 2 (page 35), we talked about using protein in regular doses in the form of snacks throughout the day, as well as in your meals, as this is the most effective way for the body to use it – and a dose before bedtime can prove particularly powerful.

The longest period over the course of 24 hours that our protein balance is in the negative is overnight, which leads to muscle breakdown, as there are insufficient amino-acid building blocks available. A leading protein metabolism lab in the Netherlands, led by Professor Luc van Loon, recently found that when 40 g of protein was ingested *following training and before sleep*, it increased muscle protein synthesis rates by 22 per cent,[13] and there were also gains in strength thanks to the increase in amino-acid building blocks available to 'drip-feed' the muscle overnight.

So protein in doses of around 40 grams before bed – which is more than most of us might normally include in meals and snacks throughout the day – can drip-feed the muscles and improve your gains following hard training.

Let's put that in context. The first priority is to have enough protein in regular doses from meals and snacks every three to four hours during the day, which can be achieved with three meals and a mid-morning and -afternoon snack. That's the biggest win.

Then, if you are training hard and your goal is to increase strength and muscle mass, an additional protein dose an hour before bed may also help. The most convenient way is probably through a protein shake (either whey protein or the slower-digesting casein protein, which was used in the Netherlands research).

Sleep Aids

Although, as we've discussed, everyone is different, evidence generally tells us that around seven hours' sleep a night is the level to aim for to best support the body's physiology. There are also some non-nutritional aids you can use, which, along with some adjustments to food, can really amplify the efficacy of your Energy Plan.

Napping

Daytime **napping** has generally been somewhat frowned on, to say the least, in many working environments, but it's long been considered a useful recharging tool in sport and for people with demanding jobs such as air-traffic control, airline pilots and NASA workers, or in forward-thinking tech companies like Google. Taking a 20–30-minute nap in early to mid-afternoon (around post-lunch slump time) can offer improvements to alertness, memory and performance, and can undo some of the effects of a bad night's sleep the previous night.

Nap for any longer than this, however, and you risk entering the deeper stages of sleep, which means that you are more likely to suffer from sleep inertia – waking up groggy and disorientated. By keeping to 20–30 minutes, you should enjoy a welcome boost without that groggy feeling, and by keeping it to the mid-afternoon, rather than early evening, it shouldn't interfere with your sleep at night.

Caffeine Naps

Now let's look at how nutrition can make an intervention and take the nap into a new realm. **The caffeine nap**[14] was put to the test in a study by Loughborough University. When sleepy drivers were given 200 mg of caffeine and then a 15-minute nap before a monotonous two-hour drive in a simulator, their caffeine nap eliminated all trace of sleepiness.

The term 'caffeine nap' might sound paradoxical, but caffeine actually takes 20 minutes or more to start taking effect, so during that window sleep is possible and, thanks to the nap clearing some of the **adenosine** in the brain (caffeine works by blocking this chemical – see page 94), you'll benefit from the boost delivered by both the caffeine and the nap itself.

If you want to try this, you need to make sure you drink your coffee briskly so it doesn't start taking effect before you start your nap. A typical coffee shop's double espresso offers a caffeine dose of 150 mg or more, and is ideal as it can be consumed quickly. Two capsule coffees from a home machine contain around 120 mg. Refer back to Chapter 5 'The Fuel Injection: Caffeine' for more details on different kinds of coffee.

But what about drinking coffee later in the day? Is that espresso after dinner in a restaurant a good idea? It really comes back to context once again.

Let's take a look at footballers preparing for an evening kick-off. About 45 minutes to an hour before kick-off, some players will strategically take between 1 and 3 mg per kg body weight of caffeine, so that it peaks in the blood for the start of the match. So for a player weighing 76 kg that would be a dose of between 76 and 228 mg – roughly the equivalent of between a single and triple espresso.

This is, of course, a very individual strategy that requires practice, but the point is that the caffeine is being used to deliver a return on the performance. In sport **everything works back from the performance itself**, and caffeine is a powerful agent that, timed well, can enhance performance. Unfortunately, as any performer will tell you, the caffeine combined with the adrenaline of the event can contribute to a poorer quality of sleep afterwards, and where possible we try to give some allowance the next day, such as a later start at the training ground, so the athlete can get back on track.

So if you're wondering about that evening espresso, ask yourself if you're prepared to take the downsides that come with it. If you're having a great time with friends, need a little pick-me-up and it's Saturday tomorrow anyway and you can enjoy a lie-in, then why not? But if you have an early start maybe just ask for the bill, please.

Extended Sleep

Simply getting more sleep in the lead-up into an event could be your answer, particularly if you are regularly getting less than seven hours a night. In one study the mood and physical performance of basketball players was found to be improved if they enjoyed extended periods of sleep:[15] the basketball players hit more free throws and improved their sprint performance. However, it's important to note that the basketball players slept for as long as possible, which probably isn't achievable for most of us, unless you plan on clearing your calendar leading up to an event.

So what about, after a hard week, simply catching up at the weekend? If lie-ins are an option for you then having one on a Saturday or Sunday after a gruelling week at work is very tempting, but is it a sustainable strategy to keep your Energy Plan on track?

There is some recent research suggesting that weekend sleep can make up for restricted sleep during the week,[16] with extended weekend sleep appearing to reduce the risk of increased mortality that constantly restricted sleep would offer.

But don't take this as a green light to go as hard as you like in the week on the assumption that you can just catch up at the weekend. This isn't going to build a sustainable Energy Plan for you, and you might also run the risk of something called **social jet lag**.[17] If you stay up later on a Friday or Saturday and then sleep in later, you're effectively living in a different time zone at the weekend; you then have to adjust once again to your weekday hours on Sunday night and Monday morning, which has an impact on your body clock and can lead to a lack of energy and low mood. It's even been linked to heart disease.

Take a Break

What did the France football team do after winning the 2018 World Cup? They went on holiday. What did Angelique Kerber do after winning Wimbledon in 2018? The same. They need some time to rest and recover (and maybe even indulge in a few off-Plan rewards).

And even on a smaller scale than post-World Cup or Wimbledon, if athletes have a particularly busy week, recovery time is built in for them to adapt (such as extended sleep after an international match or a day off after a match or hard training week). I encourage you to do the same – if you've turned up the stress to meet a deadline one week, you need to build in time to rest and recover afterwards. It should be set in stone in your calendar, an appointment you have to keep, just like your lunch and training times, as we did in our monitoring in Chapter 8.

Getting Ready to Recharge

When it comes to your Energy Plan and sleep, it's vital to get the fundamentals in place first. You're no doubt very familiar with these by now: fuelling for your demands, recovering from training and ensuring there aren't any deficiencies in your nutrition. Only once you have this in place, with a balance between stress and recovery and consistently good sleep hygiene practices, should you think about trialling any of the options on the nutrition list in this chapter to boost your sleep. Make sure you monitor your progress to see what is making the difference for you.

And most of all, make time for your sleep. Of course this largely means at night: unless you're the Tom Ford type, achieving peak performance on less than five hours a night (and very few of us are), give yourself the opportunity to have the quantity you need. But also dedicate some time to it during the day, whether that's in the form of naps or looking at how you can make a positive impact on the quality of your sleep and noticing the changes that are effective. Because while it is easy to take sleep for granted, with just a bit of self-awareness and change to your habits you could be getting out of bed with a spring in your step every morning and enjoying the abundant rewards that recharging as an integral part of your Energy Plan offers.

ENERGY EXPRESS: BOOST YOUR SLEEP AND REST WITH THESE QUICK TIPS

- Aim for a minimum of seven hours' sleep every night.
- During busy periods consider ways to supplement your nightly sleep (sleep extension, napping, catch-up sleep).
- Minimise potential nutrition inhibitors: alcohol, calorie deficit, high-fat meals.
- Trial potential nutrition enhancers: protein, carbo-hydrate, tryptophan and melatonin foods.
- Make good nutrition and sleep hygiene the foundation of your sleep routine every night.
- Consider using protein after hard training to support overnight recovery.

CHAPTER 12

Immunity

On singing star Adele's 25 tour she performed 121 live shows over a two-year period, but had to cancel her final two Wembley shows due to illness, disappointing some 200,000 fans in the process and costing the tour dearly in terms of refunds. Illness and injury on the international stage can mean millions of lost pounds in revenue, and with gruelling tour schedules this can happen to the best.

In sport, one bout of illness can have similar financial costs; for a top athlete it might mean missing or underperforming in that elusive Olympic Games final that they have been working towards every day for the past four years.

Nothing derails your Energy Plan quite like a bout of illness. You can have all the infrastructure you like in place, a well-stocked cupboard with nutrient-rich ingredients, your training plan for the week ahead and the composition of your meals worked out to meet the demands for each day... but then you get a heavy cough or cold and it all goes out the window.

The stakes can often be just as high in our own lives as they are for top performers: we work longer hours for that looming project deadline at work while trying to get in some training as well as being present at home. We find ourselves falling ill and falling behind, failing to deliver our best in every aspect of our lives and then playing catch-up as we try to recover and still meet the demands we face. Life unfortunately doesn't go on hiatus while you recover from illness.

Upper-respiratory infections (URIs), such as cold and flu, are the most common cause of illness in both athletes and the general public. Everyone is different, but we typically experience between two and four episodes of respiratory illness a year, most commonly, it will come as no surprise to read, in the winter months.[1]

No one likes to get sick, and no one wants illness to derail their plans or projects, so I'm going to share with you some of the methods we use in different elite performance settings to ensure that our athletes and performers have robust immune systems and can deliver when it's all on the line.

But first, let's once again take a look under the bonnet of these high-performance vehicles of ours to see how our bodies fight infection, so that we can get a better sense of what exactly we are doing when we boost our immune system with the Energy Plan.

Lines of Defence

The body's immune system has two main parts: the **innate** immune system and the **acquired** immune system.

The body's first line of defence is the innate immune system. These are the troops on the ground, if you like, on the front line and with you from birth. The innate immune system deals with any *general* threat, defined as pathogens not belonging to the body.

These troops seek out and destroy any clear and present dangers to your body in the form of invading pathogens (bacteria and viruses). The innate immune system includes physical barriers like skin, eyelashes and body hair, bodily fluids like mucus and saliva, which contain enzymes containing antimicrobial properties to kill pathogens, and also a host of white blood cells called phagocytes, which circulate the blood and gobble up any invaders, Pac-Man-style, to keep your body safe.

If the invader has managed to breach your front line, your acquired immune system jumps into action, although it's not actually quite a jump – this system is more complex and slower to respond than your rapid-response front line of defence. This second line of defence is more like an intelligence service, consisting of a network of cells called lymphocytes that not only attack invaders, but also collect intel on them, building a memory of them. It is this feature that is called into action when the 'friendly fire' of immunisation jabs from your GP enters the body. As we get older this intelligence service collects, through our ongoing exposure to diseases and immunisation, more data and more memories to enable it to wipe out the threat from viruses and bacteria we've experienced before.

Inviting Attack

Despite our body's multifaceted lines of defence against illness, there are unfortunately plenty of enemies within (and without) that can undermine them and lower our immunity, as shown in the diagram overleaf. Often a period of illness is down to a number of these factors clustering together to leave us exposed: travelling for a business-critical meeting, pulling an all-nighter before a day at work or an exam or even just pushing ourselves into a negative energy balance without taking the appropriate measures can all end in trouble. Let's look at each of these factors (highlighted in the illustration overleaf), so that we can see how best to prepare ourselves.

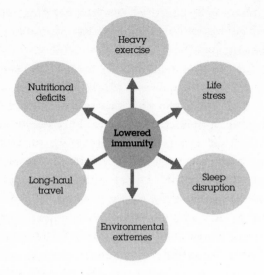

Factors that can lower immunity, adapted from Walsh, 2018

Heavy Exercise: The Open Window

Exercise, that staple salve for the body and mind and a vital part of the Energy Plan for cardiovascular health, weight management and for activating and maintaining muscle mass, can be both an ally and an enemy when it comes to your immunity.

Moderate exercise has been shown to boost the immune system and reduce the risk of infection, but prolonged, hard training causes a temporary depression of both the innate and acquired immune system over the subsequent 24 hours. This can be amplified during an intense training block that includes sessions of over an hour and a half of moderate to high intensity (55 to 75 per cent of VO_2 max). Training without eating beforehand ('training low') can also depress our immunity.

Hard, prolonged training or exercise causes a depression in white blood cell function, which can increase the risk of infection by invading viruses or bacteria. This is often termed by exercise immunologists as the 'open window' – a short-term increase in susceptibility to infection after training. It is therefore important that there is adequate time to recover between hard training sessions and that, as we'll come to shortly, both pre- and post-training nutrition fits with the demands we discussed in the Performance Plates chapter.

In an elite sports environment, the team or athlete is usually most at risk when the training or competition volume is high, the same for performing artists when on tour, partly because this is often coupled with travel, stress and even changes in the weather. For example, I have had situations at Arsenal when we have played a match in Croatia against Dinamo Zagreb on a Wednesday evening in November, followed by a lunchtime kick-off against Chelsea in London just 62 hours later. The science tells us that the body requires 72 hours to fully recover from a match,[3] yet this pattern of two or three fixtures a week can repeat itself over a month, placing players at greater risk of illness and injury. So let's take a look at what we can do with our nutrition to combat this.

Nutrition: Boosts and Deficiencies

When it comes to nutrition and boosting our immune system, we are all constantly on the hunt for the latest miracle foods and supplements. Whether it's echinacea, vitamin C, chicken soup or the proverbial apple a day, the hunt for the latest super-healer goes on seemingly endlessly. If you want to use nutrition to stay healthy while those around you are dropping like flies, your Energy Plan should address two aspects: it should ensure there are no nutrient deficiencies in your Energy Plan (such as an energy deficit for fat loss); and it should include the foods and supplements that can boost immunity.

As ever, it is vital that your fuel for each day is composed of the necessary levels of **energy** (carbohydrate), **maintenance** (protein) and **protection** (micronutrients and healthy fats).

And it is the **protection foods**, in the form of micronutrients, that we particularly focus on in this chapter.

At times when training or exercise volume is high, your priority should be to include more of the micronutrients iron, zinc and vitamins A, D, E, B6 and B12. In elite football the club chef will make a variety of different juices before training and during treatment to boost intakes of protection foods, and add options in the restaurant for lunch or snacks for players to take home.

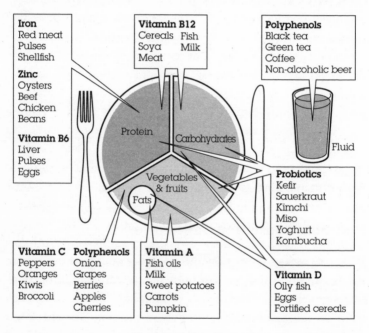

The immunity plate

Keep the Carbs Pre-training

Carbohydrate acts as a fuel for the immune system, reducing the response from stress hormones such as adrenaline and cortisol during training. Restricting carbohydrate **before** hard, prolonged training – training low – can reduce immune function.[4]

Rehydrate

During training, we usually secrete less saliva – it's what gives us that familiar dry-mouth feeling when exercising. But saliva contains antimicrobial proteins that are an important part of your innate immunity – those front-line troops once again – so you need regular fluid intake to prevent dehydration.[5]

Recover Post-training

A high-protein diet as part of recovery **after** intense exercise has been found to reduce instances of respiratory-tract illness in athletes.[6] It is also important to add carbohydrate as part of a fuelling **performance plate** or snack, especially after hard training sessions.

Boost Immunity

It is in the powerful immunity-boosting powers of micro-nutrients that many of us look to nutrition to save us from sickness, and it would not surprise me if many of you skipped straight to this section looking for a nutritional magic wand. Which might be expecting a little *too* much... However, while they aren't quite magic, there are some pretty potent micro-nutrients that set scientific pulses racing in the study of our immune system.

Probiotics

Probiotics, or 'good' bacteria, can be consumed to improve our gut bacteria and enhance our mucosal immune system, part of our first line of defence.

They are available as supplements but are perhaps most commonly associated with certain types of yoghurt.

Probiotics are certainly worth considering for those who get ill regularly, particularly with GI (gastrointestinal) issues, and they can be an invaluable part of a strategy for frequent flyers, who run the risk of travellers' diarrhoea (which we'll cover in the next chapter).

Probiotic products often contain different active strains. Lactobacillus probiotics at a minimum dose of 10 billion each day (this is the strain and dose to look for on the label) have been shown to reduce instances of URIs and sick days in athletes and non-athletes alike.[7]

> Before purchasing any supplements be sure to consult the supplements chapter, which has advice on how to source safe and reliable products.

Zinc

Zinc is a trace element that is often paired with vitamin C (albeit in lower doses) in many supplements, and is also found in many sports foods like bars and recovery drinks, as well as in multivitamins. Research has found that taking lozenges containing 75 mg of zinc within 24 hours of the onset of symptoms can decrease the duration of colds.[8] This will often mean taking a few lozenges, depending on their strength, to meet the required dose. Zinc acetate lozenges are the most effective type. Always check the label to ensure the correct dose.

Other than when you're trying to shorten the duration of a cold, you're probably getting enough zinc through your diet. Large doses may reduce the absorption of copper, another micronutrient that also supports immune function.

Antioxidants

Antioxidants play a crucial role in keeping our cells in balance and protecting them from harmful free radicals, which can cause damage to our cells. Free radicals (also called reactive oxygen species) are atoms created during energy production within contracting muscles. Small amounts act as an important signal for muscle function, but large amounts caused by prolonged, harder training can cause damage within the muscle cell, either to the DNA or to the outer layer, the cell membrane.

One of the most well-known antioxidants is vitamin C, but polyphenols are emerging as new players to consider too. Let's look at them both in more detail.

Vitamin C

How many of us are apt to reach for the honey and lemon tea or a Lemsip when the common cold strikes? Vitamin C has long been considered the go-to nutritional cure when a sore throat and sniffy nose come on, but in fact in recent years, the evidence to support its use has veered one way then another.

Cochrane, an independent global network that reviews the best evidence available to provide healthcare choices for everyone from medical professionals to patients, states that in members of the ordinary population (i.e. not elite athletes), taking 200 mg or more per day of vitamin C had no effect on reducing the number of colds. However, taking it did have a 'modest but consistent' effect in reducing the duration of symptoms.[9]

Where it gets really interesting though is with people going through periods of extreme physical stress – including marathon runners and skiers. In these subjects, vitamin C **halved** the risk of the common cold.[10]

So, while vitamin C supplementation might not be effective every day under normal circumstances, it could be effective during periods of hard training. However, before reaching for the tablets, remember that over 1,000 mg per day – a not uncommon amount in a vitamin C tablet – taken longer term (for more than three weeks) can dampen your muscles' adaptation to training. So in the first instance look to get this vitamin from your diet, using some of the sources in the diagram on page 194. And if you are going through a period of extreme physical stress, use the Supplementation process on page 237 to see if it's worth taking it for a short time.

Polyphenols

Polyphenols are found mainly in plant-based food and drink such as fruit juices, green tea, coffee, non-alcoholic beer, red wine, green leafy vegetables, red grapes, cherries, apples, dark chocolate and dry legumes. Polyphenols have been shown to have strong anti-inflammatory and antioxidant properties, and are an important tool for the muscles' recovery from bouts of heavy training. Consuming polyphenol-rich foods and drinks may also reduce the risk of respiratory-tract illness. A recent study showed regular ingestion of polyphenols via drinking non-alcoholic beer prior to and after a marathon significantly reduced the incidence of respiratory illness symptoms.[11] This provides sound rationale to include polyphenol-rich foods with meals and snacks to support your training programme, especially during heavy phases.

Quercetin, part of a subgroup of polyphenols, is found in foods including apples, peppers, red onions and kale. A study on active male cyclists found that when they took a supplement of 1000 mg day, split into two 500 mg doses for two weeks, they were less likely to develop upper-respiratory infections after hard training.[12] While that might not be enough reason for the average person to go out and purchase quercetin supplements, it does suggest that it might be worth including more of these foods in your Energy Plan during heavy stress caused by training, work or anything else.

Foods high in polyphenols:

Apples
Grapes
Cherries
Blueberries
Blackberries
Raspberries
Rye bread
Soybeans
Black beans
Hazelnuts
Green tea
Black tea
Coffee
Red wine
Cocoa
Dark chocolate (high cocoa content)
Broccoli
Shallots
Onions
Green olives
Oregano
Rosemary
Cinnamon

KEEPING IT IN CONTEXT

It's crucial to remember that these micronutrients, in supplement form at least, aren't a shopping list of essentials. It's more important to make sure you are following your Energy Plan basics really well: fuelling to meet the day's demands, enough carbs and protein around training and, finally, ensuring that there is a wide variety of fruit and vegetables in your diet to provide as many micronutrients as possible.

There are also different factors that can promote the absorption of nutrients from vegetables. Tannins and phytates are naturally found in tea and coffee, and are known as inhibitors as they can reduce iron absorption. Anyone concerned about iron levels may want to limit their tea and coffee consumption to away from mealtimes.

Vitamin C on the other hand is well known as an enhancer, and increases the absorption of iron from plant-based sources. In addition to this, adding oil to salads can promote the absorption of antioxidants (called carotenoids) from vegetables.[13]

I want to emphasise a **food-first approach** to nutrition. Cochrane noted that most supplements claiming to support immunity draw their evidence from low-quality research studies.[14] As we'll see in the Supplements chapter, many conflicts of interest exist within this sector, so the claims of many supplements should be taken with a pinch of salt.

Stress

You have relationship problems or issues at work, money's tight, exams are looming or you're moving house (or maybe you're unfortunate enough to be going through a few of these at the same time). They all add up to the same thing: stress. And this can have a negative effect on your immunity. In particular, stress has been shown to have an effect on our susceptibility to catching a cold.[15]

We can split stress into two types: **eustress** and **distress**. Eustress is essentially 'good' stress, and has been shown to have a positive impact on immunity, while distress is the 'bad' type. Whether a stress is good or bad is based around our *perception* of a stressor, which is why doing a presentation at work can make some people excited about the challenge ahead – the butterflies in their stomach being of help to them – while others might experience those butterflies as feeling sick to their stomach at the prospect.

Even the more extreme manifestation of stress, the fight-or-flight response, can be beneficial in small short doses – that is, over only a short period of time – provided there is the opportunity for the body to recover afterwards. Constantly activating the fight-or-flight response in chronic stress going on for weeks or months means we are producing the stress hormones cortisol and adrenaline beyond our normal needs. Adrenaline raises the heart rate and blood pressure, while cortisol helps regulate blood glucose. When we experience stress, cortisol elevates blood glucose to provide an energy source for the muscles and simultaneously inhibits insulin production. This results in elevated blood glucose, which can lead to increased fat storage.

Life has a habit of throwing sources of stress at us, and we can learn something from sportsmen and women in their approach to challenges and the way they frame their language around stress. You'd be hard-pressed to find a sports professional who, on the eve of a tough final or after a bad performance, would cave in and admit how stressed they are. They'll talk instead about what an 'exciting challenge' it is, how 'it was a learning experience that they took a lot of positives from'.

We can work on reframing things similarly in our own lives, whether challenges at work or in family life, and perhaps help support our immune system in the process.

It's well understood that your dose of exercise – whether this is in the form of a run, lifting weights in the gym or a yoga class – can help to reduce stress by stimulating endorphins and reducing the stress hormones adrenaline and cortisol.

We also know that emotional (or stress) eating is common when stress levels are high. Suddenly you are reaching for snacks you otherwise wouldn't, for comfort. Understanding these triggers is an important step, which is where the 'check-in' is vital. Ask yourself how hungry you are at mealtimes, and record your wellness scores (how you've been feeling generally over the last week). You can then take control and plan the next week (managing the stressors) and also ensure you're purchasing the right snacks (and removing the temptations).

Environmental Factors

Colds are a virus, we're told, and catching one by going out without warm enough clothes on is often dismissed as an old wives' tale. But, although we get colds at warmer times of year too, some recent research seems to suggest that colder air and the cooling of different parts of the body (such as nose, upper airways and feet) could be a factor. Also, and quite an obvious one, wearing appropriate clothing when training in cold conditions in the winter months. Exercise immunologist Professor Neil Walsh recently published an excellent review paper featuring some key recommendations for maintaining immune health in athletes. He highlights that the human rhinovirus, the leading cause of the common cold, is able to replicate more easily at the colder temperatures inside the nasal cavity (33–35°C) than at body temperature (37° C) because our immune systems don't operate as effectively below body temperature. Walsh advises that 'athletes are recommended to take extra precautions to avoid breathing large volumes of cold, dry air when training and competing in the winter.' This means wearing appropriate clothing, and warming up and down indoors.

Sleep Issues

We've already discussed sleep's impact on metabolism and performance, but it can also play a role in immunity. Getting less than six hours' sleep, with poor sleep efficiency (see page 175), in the weeks before exposure to the common-cold-causing rhinovirus is associated with a lower resistance to catching one.[16]

So addressing your sleep, in particular allowing yourself enough time to get a sufficient amount (at least seven hours is generally recommended, though it depends on the individual), is essential. This applies no matter how packed your schedule is in the cold winter months and particularly if you're training hard. Don't let your schedule allow you to take sleep for granted – it might come back to bite you. See the Recharging chapter (page 173) for more advice.

Air Travel

Long-haul air travel has been shown to increase the risk of upper-respiratory infection in athletes, so avoid a prolonged, heavy training session before a long flight.

Travellers' diarrhoea is a big problem when travelling to places like northern Africa, Latin America, the Middle East and Southeast Asia. Nothing saps your energy quite like a bout of diarrhoea, and it poses problems such as staying hydrated. Check out the Travel chapter (page 207) for tips on dealing with this.

Keep It to Yourself, Thank You

The pre-match handshake is a ritual central to the beautiful game, and watching the eleven players on each team shake hands with one another and then the referee and other officials is a beacon of great sportsmanship and the spirit we all hope the game will be conducted in.

It's also a germophobe's worst nightmare.

Olympic athletes approaching the Olympic Games and footballers approaching a World Cup, both tournaments that come around every four years, can be some of the worst germophobes around, desperate to avoid handshakes with the general public – and contact with anyone else for that matter – and with good reason. If someone's harbouring something and they've coughed politely into their hand only moments before, they're likely to spread what they've got. And if an athlete has been training hard, they may be susceptible to this 'open window' to infection.

The same rules apply to the rest of us. If you're competing in an event at the weekend or have a particularly tough period in your schedule and you're at a work or social event involving shaking countless hands, you're putting yourself at the same risk. Just as you are when travelling on public transport, exposing yourself to a sneeze in a packed carriage or holding the handrail and then there are public gyms with their shared equipment!

All of which could drive you mad if you let it, and your Energy Plan certainly shouldn't be an opportunity to develop a pathological fear of germs. But it is worth exercising a little bit of common sense. If you're training hard and travelling on the train daily, why not carry a pocket-sized hand sanitiser with you to apply when you get off the train? Or just wash your hands thoroughly when you get to work and when you arrive home. In the gym, try not to touch your face too much if you can help it. Your post-gym shower will be effective against germs too.

Within elite sport limiting the transmission of infections is so important to us, especially in team sports, that we look at different ways to set up little 'nudges' at the training ground to get players into good habits. One of these nudges involved attaching a container of alcohol gel to the door handle of each door at the training ground, so that when you squeezed the handle to open it, your hand was given a dose. Reading this, it may well sound a bit extreme, but you might feel differently if you found yourself at your own starting line (metaphorical or literal) with a runny nose and a tickly cough.

ENERGY EXPRESS: THE IMMUNITY PLAN

- **Follow good hygiene practices:** Hand-washing, carrying portable hand sanitiser, good food hygiene when travelling.
- **Limit environmental stressors:** Avoid situations involving contact with infected people, especially in the winter months.
- **Manage training volume:** Modify hard and prolonged training sessions if feeling unwell. Plan easier training sessions the day after harder sessions. The 'below the neck' rule is a common recommendation: if symptoms are above the neck only (runny nose, sore throat), then you can train, but if below the neck (chest congestion, aches, diarrhoea), you don't.
- **Prioritise your sleep:** Ensure you are hitting seven hours a night.
- **Fuel and recover appropriately from training:** Ensure you aren't energy-deficient. Training low increases stress on immune function. Use carbohydrate and protein to refuel and repair after hard training.
- **Add foods rich in the following components:** Vitamin C, probiotics, polyphenols, vitamin D and zinc. These may boost immunity in given scenarios.

CHAPTER 13

Travel

Tom Stuker is a real high-flyer.[1] In March 2018 Stuker, an American car-dealership consultant, reached 19 million air miles, accrued over the best part of 40 years, and, given that he racks them up at a rate of over half a million per year, he's likely on his way to twenty million by the time you read this. He spends 15 days a month travelling, making journeys from Newark in New Jersey to Sydney, Australia, which accounts for nearly 20,000 miles per round trip, and his total mileage would circle the Earth (at time of writing) 763 times. His peripatetic existence hasn't gone unrewarded, either. United, the airline with which he has accrued his miles, named a plane after him after he hit ten million, and as a thank you arranged for him to throw the first pitch at a Major League baseball game.

Stuker is about the most extreme example you will find, but some degree of travel is a factor in most of our lives – performing artists, professional sportsmen and women and the rest of us. For example, during the 2017 season golfer Paul Dunne spent almost 378 hours either in a plane or at an airport, some 16 days of his life,[2] but the challenge – for him or for any of us, even if only flying to an event once a year – is how to arrive at your destination and be on form so that you can deliver a great performance.

Due to the fine margins in professional sport and the resources at our disposal, we plan travel schedules to the finest detail. In elite football the plane will often be a charter flight, specifically for use by the club. Menus, and the timings of food served on the plane, will be to the minute, training sessions for once the players arrive will be planned in advance to help them adapt quickly to the new time zone and, finally, the chef will travel with the team to ensure that the food in their hotels, in whichever part of the world, is delivered to the correct specifications.

Of course, most of us can't exercise this kind of extreme control; air travel can mean catching a 7am flight, which means getting up at 4am or even earlier to get to the airport. We're often required to arrive at our destination, give our presentation or whatever the work is and then return the same day – and are expected to get on with business as usual at work the next day. And even if it's not work but a mini-break or romantic getaway you're catching that early flight for, it can often mean getting there completely exhausted and writing off the first day.

As far as your Energy Plan is concerned, there are two types of travel – **performance** and **pleasure**. Performance travel might be a trip to an event for work or a meeting abroad; competing in an event such as the New York City Marathon; or a physical challenge such as a hiking holiday along the Inca trail in Peru or a cycling holiday in Majorca. Whatever your purpose, it's all about needing to get to your destination and being able to deliver the goods, despite the potential derailment of travel.

And by pleasure travel I mean primarily holidays for which getting there ready to deliver might not quite be so important, such as a relaxed beach holiday.

We'll take a quick look at pleasure travel first, as the bulk of this chapter will be devoted to performance travel; even if you only ever travel abroad for holidays, it's still worth reading the performance travel section for advice that might work for you. After all, if you only have a couple of days away on a city break and you're taking an early-morning flight to get there, you won't want to waste a whole day feeling shattered.

Pleasure Travel

For many of us, there's nothing quite like that feeling of arriving at the airport ready for a holiday. No need to think about work for a week or two, some reliably warm weather to come, the tempting food vendors at the airport – and don't those chilled glasses of foaming beer or fizzing prosecco at the airport bar look like an appealing way to toast the beginning of a particularly exciting trip?

With your Energy Plan firmly in place at home, it's easy to think that, because you've been going at 100mph with your activity and nutrition, you'd be justified in pulling up the handbrake entirely for a week or two away. But you might just want to change down a gear or two first…

Our athletes typically have around three to four weeks off in the summer to decompress and unwind, and they will typically go on holiday to do this – they're no different from the rest of us in that regard. But they are aware that over this time there still needs to be some structure, as muscle mass and strength can reduce during the off season.

Studies have been conducted on the impact of two weeks of increased sedentary behaviour – like lying on a sunlounger for two weeks – and have found that, in active people, it causes 'metabolic derangements' in the form of reduced fitness and insulin sensitivity, and has a negative impact on body composition.[3] These effects, alarming-sounding though they are, are reversible and in these studies did reverse themselves once the subjects returned to normal activity levels (thankfully). However, the results weren't so uplifting in another study that looked at older, overweight and pre-diabetic people who were less active; it found that the reduced muscle protein synthesis and insulin sensitivity caused by their inactivity did not return to pre-holiday' levels.[4]

I am not for a second suggesting that you shouldn't put the stricter parts of your Plan to one side to take a break, get some rest and indulge in some off-Plan rewards. What else is a holiday for? But try to follow the lead of the elite athletes I work with. They make sure they get enough maintenance foods in their meals, so that they aren't just bingeing on carbs, and they make the effort to keep themselves moving, whether that's by going for a long walk or a run, or swimming in the sea.

It's not so hard to incorporate this sort of activity into your own holiday without compromising your enjoyment. If you have a pool where you're staying, or you're near a beach, get into the water for a swim as much as you can. Go for a walk along the beach; or, if you're a keen runner, there are few more enjoyable routes for your run than a beach at sunrise or sunset. If you have a young family, join in with the kids to keep active: Frisbee, the ubiquitous beachside bat and ball and volleyball are all ways of keeping you moving that feel like a 'holiday' from your usual training routine.

Keep this basic balance and, like our football players returning to training at the start of the season, you might not quite be in the shape you were in before you left, but it won't take long to get your Plan back on track – while still enjoying a great holiday. And remember, you'll have the contingencies to get yourself back on track that we discussed in the Winning Behaviours chapter (page 115).

Performance Travel

For the rest of this chapter, we're going to look at dealing with travel as a part of your Energy Plan that you can manipulate so that you're better able to deliver a performance once at your destination. Let's start by taking a look at the biggest problem long-distance travel poses to your Energy Plan.

Jet Lag

Finding yourself wide awake at 3am in a new time zone after flying long-haul, while it feels like the whole world around you is sleeping soundly, must be among the most lonely and frustrating feelings there is. Jet lag strikes indiscriminately, and comes with a host of unwelcome symptoms like fatigue, low mood and problems with the gut, such as constipation, lack of appetite or diarrhoea.

Jet lag is caused by moving across multiple time zones in a shorter period of time than we are able to process. Our body clocks are primarily set by the light–dark schedule of our environment through melatonin, otherwise known as the 'vampire' hormone because it is produced when it gets dark to induce sleepiness. This works provided we have been awake long enough (melatonin production in the body reduces dramatically when it gets light again).

Emerging evidence suggests that the time at which we eat may also play a role in synchronising the body clock, through metabolic changes (such as blood-glucose rhythms).[5] After long-haul travel our body clocks are out of sync with the new time zone, and they need time to catch up. It might be 3am in New York, where you're staying the night, but your body clock is still on London time, where it's 8am already and you're ready to start your day.

Our body's biological clocks have to move an hour for each time zone crossed. Travelling east, it's estimated to take about a day for each time zone crossed for your body clock to sync to the local environment, while for travelling west it's more like half a day per zone.[6] So if you're travelling going east by five time zones, it can take you as long as five days for your body clock to adjust. It might even take longer, as we're all different and there's no predicting how each individual will be affected.

Preparing to Battle Jet Lag

If travelling west, the best time to arrive at your destination and adjust to jet lag is mid-afternoon, so that you get light exposure – the primary setter of your body clock. This is because you will feel sleepy earlier at your destination.

On the other hand, if you're travelling east, where you will find it harder to get to sleep at your usual time, the ideal time to arrive is at nightfall, so you can avoid daylight and only get your first light exposure the following morning. Unfortunately, flight schedules aren't tailored to meet the needs of our body clocks, but if there is a choice then book flights as near as you can to these points in the day.

Online resources such as the British Airways jet-lag advisor,[7] developed in conjunction with the Edinburgh Sleep Centre, and the Jet Lag Rooster,[8] developed with the Mayo Clinic Center for Sleep Medicine and Rush University Medical Center in the USA, are good for advice tailored to your journey. You input details like your arrival and departure locations and times, as well as some information about your sleeping habits, and they give advice on precise times of day to seek out light and when to avoid it after you land.

Travel Fatigue, Sleep Disruption and Immunity

Travel fatigue is a condition that doesn't get as much attention as jet lag, but this temporary exhaustion is a very real factor when it comes to travel – and it doesn't have to involve a plane. Any form of transport can induce travel fatigue: long road journeys as well as air travel.

Lack of comfort, prolonged sitting in cramped conditions and the increased stress involved in getting to the airport on time, or fear of flying, can all contribute to travel fatigue. With flying, when you add on the amount of time spent getting to the airport and waiting around to the duration of the actual flight, even short flights can end up being all-day-travel affairs that tire us out and leave us short of energy on arrival. And while the after-effects are short-lived, they can accumulate over periods of a lot of travel, leading to a cocktail of fatigue, sleep disruption and reduced immunity, as highlighted in the diagram below.

Both quantity and quality of sleep have been shown in studies to be reduced during international travel, which can only add to the symptoms jet lag produces. International travel can have a negative impact on immunity, and the act of travelling long-haul by air in itself produces a two- to fivefold increase in upper-respiratory infection symptoms,[9] although it isn't yet clear whether this is because of the hypoxia (reduced oxygen) or the increased exposure to pathogens (viruses and bacteria) on board.[10]

Travel and performance expert Dr Peter Fowler has conducted a large amount of research into travel and athletes, and his study results support the interplay between jet lag and travel fatigue and their effects in the diagram below.

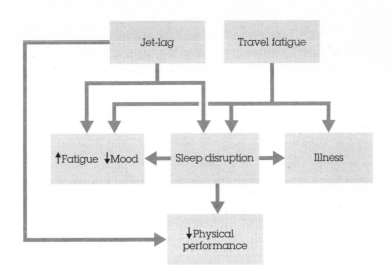

Adapted from Fowler, 2018

Travel-induced sleep disruption and illness – the impact of which we've looked at in the Recharging and Immunity chapters – combined with the increased fatigue and potential impact on mood all come together in an almost perfect storm of stressors to take their toll on us during and, particularly, after travelling.

Plotting Your Travel

The effective nutrition element of your Energy Plan involves making your meals work harder for you, delivering the fuels that your body requires according to your demands. But when we're travelling, maybe running late and feeling particularly hungry, it's easy to either miss a meal or have to make a snap decision, and so it's very easy to lose control of our Energy Plan. Our appetite hormones (including leptin and ghrelin: see page 179) respond, driving us to eat and making everything look a little more appealing – especially calorie-dense foods, high in sugar or fat.

So how can we avoid these options and exercise some control over our travel so that we arrive at our destination, if not entirely refreshed, then at least with enough energy to put in a decent performance? And also so that when we arrive home, fatigue and ill health don't follow us around for the next week?

With the football teams I've worked with, the squad would mainly be on a **high day** when we travelled (see page 109) to fuel their bodies, so snacks would include fruit, home-made flapjack or granola bars and sushi. On the way home, however, the goals were different: for those who didn't play and wanted to keep their fuel intake **low**, snacks would include sashimi and low-fat Greek yoghurt. This is more in line with what your fuel requirements are likely to be, unless you are travelling to a challenge or competition such as a marathon.

TRAVEL DAY = LOW DAY

Unless you are training on a travel day, it will be a **low day**, which means higher-protein, low-carb snacks, such as:

Low-fat Greek yoghurt

Tuna or salmon sashimi

Edamame

Spiced mixed nuts and seeds

Whey or plant protein shake

Spiced chicken or tofu bites

Turkey roll-ups

Protein bar

Low-day structure – this would need tailoring according to flight time:

Timing	Feeding
Breakfast 8-9am	Maintenance
AM Snack	Maintenance
Lunch 12-2pm	Fuelling
PM Snack	Maintenance
Dinner 7-9pm	Maintenance

In professional sport and travel, two key principles are followed as long as time allows:

- Light training session before departure
- Meal before departure

You can include these principles as part of a four-phase travel planner to meet the challenges travel poses to your Energy Plan. Some of the points in this planner apply only to long-haul air travel, some to air travel only and others are more general. Pick and choose from the information to build your own bespoke travel checklist.

1. Pre-travel (weeks and days before)

Food Availability

One of the biggest worries is being able to source the foods you'll require when away, particularly if you have specific dietary considerations such as being vegan or gluten-free. Speak to your hotel in advance and research local food shops in the area online, if possible. Also, reviews on TripAdvisor or advice from friends who have visited the area before can help you identify where the gaps or challenges will be for your trip.

Food Hygiene Issues

Travellers' diarrhoea is an unpleasant and common potential risk of long-haul (and even short-haul) travel, with over 12 million cases reported every year.[11] Areas like northern Africa, Latin America, the Middle East and Southeast Asia have a higher reported risk, with the rates of TD ranging from 5 to 50 per cent,[12] but it's worth remembering that you can catch it anywhere, even in low-risk areas.

The usual advice you'll read and hear about high-risk areas applies: being careful with undercooked meat and shellfish, avoiding raw, unpeeled fruits and vegetables, taking a pass on the side salad and staying clear of tap water, ice and non-pasteurised milk. Areas to be particularly vigilant about are street vendors and smaller farmers' markets and restaurants.

If you're very concerned about this, you can be proactive by taking a probiotic prior to and during travel. The cocktail of stress, jet lag, unfamiliar foods and water can disrupt normal bacteria and our body's natural defences, and probiotics contain 'good bacteria' to help reinforce this barrier to pathogens attaching and colonising in the intestines. See the Immunity chapter (page 189) for more detail on probiotics and which type of probiotic to take. With our athletes we recommend starting the probiotic two weeks prior to departure.

And if the dreaded travellers' diarrhoea does strike, rehydrating through drinking water, electrolyte rehydration salts and carbonated drinks, consumed little and often, are enough to deal with most cases.[13] Withholding solid foods for the short term (up to 24 hours) is also a good idea, or, if you can't manage this, a good alternative is consuming a bland 'BRAT' diet, consisting of banana, rice, apple sauce and toast, and avoiding alcohol, fat-rich foods and dairy products until the diarrhoea settles.[14]

Pack Your Essentials

It's worth having some travel essentials in your hand luggage so that your travelling won't interfere with your Energy Plan too much. Any or all of the below can help keep you on track:

- Protein sachets – individual sachets of whey or plant protein
- Protein shaker – to mix the shake
- Meal replacement drink (optional) – discussed in Supplements chapter, page 229
- Alcohol handwipes
- Eye mask, loose clothing, earplugs, neck pillow
- Snacks for plane (goal-dependent – either maintenance or fuelling snacks)
- Immune-boosting supplements (zinc acetate lozenges, vitamin C, probiotics)
- Water bottle

Be aware that items permitted in the UK could land you in hot water and with a big fine in other countries. Even though there is no hard-and-fast rule, I would recommend you ensure that all supplements (pills, powders and potions) are labelled to avoid questions and delays at customs, and that you have any paperwork for them, such as receipts or batch-testing certificates (see Supplementation chapter), with you.

Know Your Itinerary

Stressful travel days are the worst – last-minute packing, rushing to make your transfer or wolfing down a fast-food meal at the airport. Make a plan the day before, and I can't emphasis enough that this should include what and where you will eat, whether it's bringing your food with you or choosing an option at the airport that meets the needs of your Energy Plan.

2. Day of Travel

Quick session?

For our athletes a short, sharp training session before travel is always part of the plan. Factor this in if there's time; it's a good idea to get some physical activity before a long period of sitting, and it may also offer some stress relief. The main thing is to avoid long, heavy training sessions, which can reduce your immunity and increase your risk of infection.[15] Something like a jog round the park or a session on a static bike is perfect.

Don't Rush

Give yourself time to eat your preferred meal at the airport. This will allow you to avoid eating your main meal on the plane and minimise the need for unnecessary snacking.

Remember your goals

Travel days often mean less exercise, which means you need less fuel. It's easy to get swept up into a holiday mentality along with everyone around you at the airport, with all of its temptations, from fast-food outlets to a bar where you could have your first celebratory drink of the trip. Take your own snacks on board to meet your goals, and most importantly have the mindset to stay on track, especially if the culmination of that important goal is just around the corner.

3. During Travel

Get on local time

Set your watch while still on the plane, as soon as you know what the local time will be, and try to align your sleeping patterns with your destination. If it's night-time there while you're flying, sleep on the plane, using an eye mask, earplugs, a neck pillow and whatever else you require to help you get some rest. If it's morning there and you feel tired, try not to go to sleep – consider having some caffeine to keep you awake and alert during the flight. As per the TTA model on page 62, the **Timing** of your food (both on board and on arrival) is important to support your adaptation to the new environment.

The in-flight menu

The Argonne Diet and the Harvard Fast are two approaches widely discussed in the US. One involves feasting and fasting in the days before the flight, the other fasting during the flight and using a big meal on arrival to help with adaptation. While the evidence is limited, it is true that fuel requirements for the body definitely reduce during travel because even the lightest training plan would be difficult to achieve in an economy-class seat (so carbs won't be doing you any favours). My general recommendation is that your travel day is a **low day** (see the planners on page 108), unless you have trained pre-flight or, as in the case of many of our athletes, are fuel-ling for an event the following day. Although the temptation will obviously be there, it's important not to eat the snacks on offer on the flight. Bringing some protein-rich snacks on board will help you avoid making 'bad' choices.

Constipation can also be an issue on long-haul flights. Important steps to consider here are increasing fluid intake and including plenty of fibre-rich foods with your meals such as fruit and vegetables. Exercise also aids gut motility, so it's important to get active before and after your flight, as well as getting up to stretch your legs during it.

WHY DOES PLANE FOOD TASTE BAD?

For some, plane food is a guilty pleasure and all part of the holiday experience. For others, usually those who fly more frequently for work, it's to be avoided at all costs. As Michelin-starred chef and TV presenter Michel Roux Jr succinctly puts it: 'Eat before you get on the plane.'

But what exactly is the problem with plane food? Usually – certainly in economy class, at least – the food is prepared on the ground and frozen, hours before you'll actually eat it, and then heated in the sky. It's usually covered in sauce to prevent it going dry, and even if it did taste great on the ground, it's unlikely to on board. The dry air in the cabin affects our sense of smell, which changes our perception of how food tastes, while changes in air pressure and background noise on the plane repress our taste buds. Because of this extra sugar or salt, or strong seasonings, are often added to the food to give it more flavour.

If plane food is an infrequent guilty pleasure for you, then it's up to you whether you indulge. But if you're a frequent flyer, follow Michel Roux

Jr's advice and get yourself something else either before the flight or on arrival. That way your Energy Plan won't be affected as much – and your food has more chance of actually tasting of something.

Hydration

The cabin pressure and dry air will dehydrate you on board (we lose moisture as we breathe the drier cabin air at altitude), so make sure you take on plenty of fluid. Think about the type of drinks you'll take on board with you – it's worth noting that some flavour or electrolytes can promote voluntary drinking, meaning you are more likely to stay hydrated.

Refer back to 'Water' on page 41. If you're going to the toilet regularly and your urine is a light colour and plentiful, then it's a good sign that you are adequately hydrated.

Move and stretch often

Some athletes wear compression leggings when they fly. This might be a bit extreme for your needs, but compression socks are very common and widely available. However, your first priority should be loose, comfortable clothes – even clothes you feel comfortable sleeping in, if you're planning to sleep on the plane – and you should also get up and move regularly to avoid lower-limb swelling.[16]

4. Arrival

Let there be light

Increasing or restricting light exposure is the most important intervention for resetting your body clock to destination time.

If you're travelling west, seek light in the evening; travelling east, restrict light exposure if you arrive at night-time and seek it out in the morning. Restrict blue light from phones and tablets before sleep.

Ease into your routine

Combining light exposure with some exercise outdoors may help to accelerate your body clock adapting.[17] However, don't go straight into heavy training as your coordination may be impaired by jet lag. Intense training such as sprinting has been shown to be impaired in the 72 hours after long-haul travel – training should initially be lower in intensity.

Embrace that morning coffee

Morning caffeine in the first five days after arrival has been shown to improve both cognitive and physical performance.[18] So if you're a caffeine drinker, embrace your usual morning coffee (or whatever your delivery method of choice is), using your particular dose as we talked about in Chapter 5. And if you're not, getting out into the daylight when you need it is doubly important to help you adjust.

Orientate yourself

Get 'on plan' with your nutrition principles as soon as possible. Although it can be tempting to stay in your own bubble on arrival when you're tired, get out and have a look around to orientate yourself and find the local shops and restaurants that will form part of your plan. Talk to people – keep yourself active and your mind engaged.

CHAPTER 14

Supplementation: Lifting the Lid

I t was prior to the 2008 Olympic Games in Beijing, and I was sitting down with an athlete for the first time to discuss supplementation.

The athlete began putting supplement containers on the table, one by one. And on it went, to the point that one of the coaches walking past asked if we were playing chess. In the end there were on the table no fewer than 28 different supplements that were part of this athlete's current regime.

So we started from the beginning: I asked for a rationale as to why he was taking each one. He started off with the ones he considered the most important, and there were some solid responses:

'Energy on competition day.'

'To boost my immune system during the winter.'

'To help improve my strength during training.'

'As an insurance policy, in case I'm missing anything from my diet.'

But as we went down the list, the answers became more hazy:

'I'm not sure – I haven't taken that in a while.'

'My wife said this is good for me.'

'An old trainer recommended that for me.'

This meeting went on for a while. And this was before we'd even got into discussing whether the supplements were effective, safe or where they had come from (the sourcing). This last point is extremely important because athletes are governed by WADA (World Anti-Doping Agency) rules, and not only drugs, but also dietary supplements can contain banned substances that can cause an anti-doping rule violation.

With this athlete's training, everything was meticulously planned and executed, but when it came to supplementation there was no process – everything was just thrown together.

This is an extreme example, and obviously you're probably not going to be subject to random drug testing any time soon. But there are some parallels here that will apply to you if you've ever taken dietary supplements, even if that's just in the form of something like a vitamin tablet or protein shake, and particularly if you are looking into supplementation more seriously as part of a training plan.

So, just as I did with this athlete, I want you to get your supplement bottles on the table and think about why you're taking them: all of them.

Write down each supplement you have taken over the last six months, along with the reason why you took it. We will revisit this later on.

One thing I want to make absolutely clear from the start of this chapter: **there are very few supplements that actually work.**

The world of supplementation can feel a bit like the Wild West at times, awash with confusion and contradictory advice, and it's little wonder. The supplement industry is expected to be worth around 220 billion dollars by the year 2022[1] and the multinational companies behind supplements have an arsenal of seduction techniques – traditional channels such as magazine and billboard advertisements, the Instagram influencers subtly beckoning you with their ripped bodies and legion of followers to the product in their hand, even that helpful person at the health-food store or the personal trainer at your gym – to convince you that their supplements could change your life. The influencers can be even closer to home – maybe you have a friend, colleague or partner telling you about the fantastic results the product they've been using has yielded.

As many as 93 per cent of elite athletes have used supplementation in recent years,[2] and their use among the general public has been on the rise over the last decade too; dietary surveys in the USA, for example, suggest that it's currently a whopping 50 per cent.[3]

A common comment I hear is 'But they're only supplements – what harm is taking some extra vitamins going to do?' It's easy to fall into the trap of thinking that because something is good for us – or, with something like vitamin C, *necessary* to us – it's a case of the more the better. We've seen this before with so-called 'good' foods – have a dollop of coconut oil here and a handful of nuts there, wash it all down with a kombucha drink in the misguided belief that this will make us more healthy.

With supplementation, just as with any type of food you consume, often **the dose makes the poison**. Mega-dosing (high doses well above the RDA) is common, as we often fall into the trap of thinking a little bit more will yield even more benefit, forgetting that fact that supplements can have powerful effects, not just positive but negative too.

In fact, it has been shown that large doses of vitamin C (1,000 mg per day and upwards) and vitamin E (235 mg per day and upwards) may negatively impact on how the muscles' mitochondria – those little power generators in your cells that convert fuel to your body's preferred energy currency – adapt to endurance training, potentially dampening the effect of your exercise programme and interfering with your Energy Plan.[4]

And while it's well known that iron deficiency and anaemia impair athletic performance, so too does too much iron; supplementation when iron levels are already sufficient can cause iron overload, highlighted by blood markers in up to one in six recreational male runners.[5] We shouldn't be using supplements if levels are already sufficient, and the only way to be sure of your levels is with a blood test. So exercise caution.

However, when we've been training hard, working hard and trying to fit in a social life around it we can feel like we're in need of some extra help for another big day tomorrow – so when should you consider supplementation?

Rocks, Pebbles and Sand

Stephen Covey, author of *The Seven Habits of Highly Effective People*, talks about getting what he calls the 'big rocks' – the most important priorities – in place first. And we can borrow this approach when we look at supplementation. Instead of immediately turning to a supplement, we should look at our big rocks, those key elements that we have been discussing throughout the Energy Plan, which are:

1. **Fuelling**: the right amount of fuel for energy (carbohydrate) - during training and working days - for your energy balance
2. **Recovery**: the right type and timing and amounts of maintenance foods (protein) to support muscle growth and repair
3. **No Deficiency**: all of the protection foods (micronutrients) to support health and performance

Next to these, supplements are simply pebbles and sand. Used appropriately, yes, they can play an important (if small) part in your Energy Plan, but it's the big rocks that will make the more meaningful difference to your results and life, and it's pointless getting started with supplements without having those in place first. As emphasised throughout this book, the Energy Plan's approach to nutrition is a **food-first** one, as this is where we can make the biggest gains.

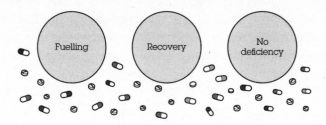

The 'big rocks' of the Energy Plan

Supplements Revisited

At the start of this chapter I asked what supplements you'd taken over the last six months and why.

I see so many people whose list of supplements, just like that of the athlete we met earlier in the chapter, has built up year after year to the point that their kitchen cupboards are filled to bursting. And when I ask *why* they take each individual supplement, or if they know if they even *need* to, they can't tell me. Below are some of the most common reasons why people take supplements. Do any of them tally with what you've written down?

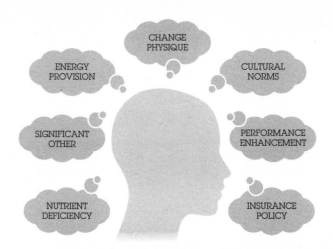

Common reasons for taking supplements

So before we address whether using a supplement is a good idea, it's important to define the supplements that actually work according to their role – why you would use them. There are three main categories of supplementation that we use in sport:

1. Sports foods

Sports drinks, gels and bars

Sports drinks, probably the most commonly used and recognised sports 'food', provide both fluid and carbohydrate during exercise. These typically contain around 6 per cent carbohydrate (so a 500 ml bottle delivers around 30 g carbohydrate, a key amount that we discussed in Chapter 3), and also sodium, which promotes fluid absorption.

Gels typically contain around 25 g carbohydrate for use during exercise, whereas confectionery such as shots or chews deliver around 5 g per piece (35–50 g per packet). These often also contain electrolytes, and sometimes caffeine.

Protein supplements

Mainly for use in post-training recovery, potentially in combination with carbohydrate (for refuelling), they are a convenient snack to meet daily protein needs or for use during travel.

Protein supplements may come in **powder** form, from animal (such as whey or casein) or plant (soy, rice or pea) sources. These typically provide between 20–50 g per serving (see protein recovery, Chapter 3, page 57).

Protein bars offer a solid alternative to the powder form, and broadly fall into two categories – low- or moderate carbohydrate.

Protein-enhanced foods are a rapidly developing area of this market. This is when protein is added to more usual foods such as yoghurts, ice cream, cereals and cereal bars.

Meal replacement drinks are used to replace meals during busy schedules and to support either high-energy intakes during weight gain, high training volumes or to replace meals as part of a diet for weight loss. These are typically either high-calorie or low-calorie (depending on their function) and provide a complete range of macro and micronutrients.

High-electrolyte products are for use in replacing large fluid and electrolyte losses from making weight, endurance events or during sickness. They typically contain added sodium and potassium to treat dehydration.

Sports foods are an extremely varied category and the composition of each of these products will vary dramatically depending on the brand, so it's important to check the label (alongside the source, as we will come on to) to ensure you have the right product.

2. Micronutrients

The two main reasons for taking micronutrients in supplement form are to correct a deficiency in a particular nutrient, and to support immunity (see the immunity chapter for more information).

Micronutrients of which people are most commonly at risk of deficiency include vitamin D, iron and calcium (see Part I, page 39). Micronutrients such as vitamin C, zinc and probiotics can help to support different elements of immune function.

3. Performance supplements

This is the final category to consider for your Energy Plan, but are only to be considered or used when the 'big rocks' are already in place. We've already discussed caffeine (see page 94); these supplements are similar in that they have a direct effect upon performance. They should be considered individually (see overleaf). You should consult a registered sports and exercise nutritionist if you're considering using any of these alongside your training programme.

Creatine

Creatine, made from three different amino acids, has been shown to improve performance in repeated bouts of high-intensity work (such as resistance training or sprinting) enhancing training capacity and adaptations (such as muscle strength and power).[6] It increases muscle **phosphocreatine** stores, which recycle ATP, the body's energy currency, for explosive movements – a far quicker process than that involving carbohydrate.

You could consider using creatine monohydrate (the most effective type) for a period of training in which you are aiming to increase strength and muscle mass, or to enhance training capacity during intermittent training or resistance training programmes.

There is potential for a 1–2 kg increase in body mass if using creatine, through water retention in the muscles, and with appropriate use it has no negative long-term effects. Creatine is found within red meat, poultry and fish, and levels are typically lower in both vegans and vegetarians,[7] which means that creatine supplementation maybe an important consideration to support training goals if you follow these diets.

Beta-Alanine

Beta-alanine[8] is a newer player on the scene. A non-essential amino acid, it potentially improves high-intensity exercise performance (30 seconds–10 minutes in duration)[9] and reduces the acidity in the muscle (the burning feeling), causing fatigue.

It can support exercise capacity during high-intensity training, so may be used for increases in strength and muscle mass, team sports performance or even high-intensity-based training programmes. Sometimes this is chosen instead of (or even used with) creatine, due to creatine's potential to cause fluid retention.

Dietary nitrate

Dietary nitrate has received a lot of press in recent years in the form of beetroot juice thanks to its capacity to affect many physiological aspects of exercise performance. It can improve endurance performance by reducing the oxygen cost of submaximal exercise, essentially making the muscles more efficient, and may also improve muscle power and sprint exercise performance.[10]

You could consider using it for endurance-based training, although there is a sound rationale for anyone undertaking hard intermittent training programmes to include more nitrate-rich vegetables as part of their daily protection foods.

A 'dose' of nitrate (5 mmol) can be provided by half a litre of concentrated beetroot juice. It's also found in green leafy vegetables such as lettuce, spinach, rocket, celery, cress and beetroot in its whole form, all of which typically contain over 4 mmol per 100 g fresh weight.[11] So that means, for one dose, eating more than 100 g of nitrate-rich vegetables. If taken 2–3 hours before training or competition, concentrated juice or shots are the most efficient, but these need to be trialled as they can cause GI upset.

Sodium bicarbonate

May also improve high-intensity exercise performance by reducing the acidity within the muscle, prolonging the time to fatigue. A supplement well known to many middle-distance athletes, it has potentially explosive effects if not well managed (it commonly causes gastrointestinal discomfort and diarrhoea). It's used in very specific training scenarios. So like all supplements, if you're going to use it, try it in training to see how you react to it.

What About the Others?

So, having looked at the supplements that work and which we use in sport, you might be thinking, OK, what others are there? After all, our athlete at the beginning of the chapter had 28 on the table. While we can't look in detail at every single supplement available, here are some guiding principles on the rest:

Supplements Claiming to Reduce Body Fat

Other than increased protein intake, there isn't anything on the market with strong evidence in this area. These products are some of the most likely to be deliberately adulterated, and if your goal is to reduce body fat, it's better to use training and food-based nutrition to achieve this.

Omega-3

An essential fatty acid. There's growing evidence of its benefits in multiple areas such as cognitive function and muscle growth and repair. Increasing intakes in the diet should be the first consideration before you turn to supplements; we looked at this in Part I.

Anti-inflammatory Supplements

The interest in anti-inflammatory nutrition is on the increase. In this area, **tart cherry juice** and **curcumin** may be beneficial in certain situations for muscle damage and inflammation – although due to their potential to dampen muscle response (adaptation), more research is required before these can be recommended.

GELATIN, VITAMIN C AND COLLAGEN

Over fifty per cent of injuries involve musculo-skeletal tissues (muscles, tendons, ligaments etc), which are rich in collagen, a crucial substance for making these tissues stronger so that they can better withstand the forces from training.

Professor Keith Baar and his team at the University of California have produced research showing that 5-15 g gelatin enriched with vitamin C, taken an hour before training, enhanced collagen synthesis.[12] (Collagen hydrolysate can be an alternative source, for those not able to use gelatin.) Gelatin is a flavourless protein made from the connective tissues of animals, commonly used in jelly, desserts, confectionery and bone broth. This exciting new area of research has seen many professional teams start to make jelly shots for their teams to take an hour before training as an injury prevention strategy.

You can replicate this as part of your own injury prevention strategy, making a jelly shot in the same way as you would make ordinary jelly using gelatin, fruit juice (a flavour of your choice) and water, and ensuring each shot contains a 50 mg hit of vitamin C.

Evidence

The claims made by supplement brands and 'health gurus' are certainly dazzling and authoritative-sounding. Expressions like 'science-backed' and 'evidence-backed' are common; on the surface this may sound reassuring, but the truth is that supporting evidence can be found for almost anything.

Highest quality evidence
Derived from analysis of multiple studies that meet strict criteria.

High quality evidence
A single research study measuring effectiveness of supplement in specific scenario.

Low quality evidence
Often the most prevalent in the public domain within blogs and articles, and on social media.

Meta-analyses and systematic reviews

Controlled trial

Other studies including cohort, case control, cross-sectional and case reports

Ideas, expert opinions, editorials and anecdote

Adapted from Maughan et al., 2018

It's useful to think of the hierarchy of scientific evidence as being a pyramid structure, as in the diagram above. At the bottom of the pyramid is the lowest level of rigour – anecdotal evidence, the thoughts that appear in some online articles and ideas that this *might work*. At the pyramid's peak are meta-analysis and systematic reviews, which review all of the available studies that meet the strict inclusion criteria, systematically analysing whether a supplement works and the circumstances in which it is best applied. This is the gold standard in which the available evidence from studies that meet the right criteria are used.

Unfortunately, many 'science- and evidence-backed' claims on products come from the bottom rung of the pyramid, which is the kind of unregulated advice widely available in any blog or magazine and is no guarantee that the supplement is effective. So next time you read or hear something in the media about supplements and 'evidence', it's important that you assess whether that evidence is from a credible source.

Most people will read about studies in secondary sources like newspapers or online news outlets. You can easily find out whether a study was a systematic review or something further down the pyramid: often in online articles there is a link you can click to take you to the source; or search for it online by putting in a few key words and names from the story you're reading.

Risks

Supplement manufacturing isn't as tightly regulated as the pharmaceutical industry. In fact, in most countries supplements are regulated in the same way as food ingredients: there are no requirements to prove their claimed benefits or safety or any quality assurance of content.[13]

While for many people and products the worst might be that the supplement simply does nothing, there are some very real health risks. In the US an estimated 23,000 A&E visits each year are associated with dietary supplements,[14] and there have been numerous reported cases of liver toxicity, cardiovascular problems, seizure and even death connected to weight-loss supplements.[15] And it's an international issue. In a number of cases, the problem has been caused by an ingredient in a product not declared on the ingredients list.

There is currently a real disconnect between this area and that of awareness of the food we consume. We live in an age of lively engagement with food provenance and quality: we want to know where our food is grown or reared and the method of its production. We want to know about its journey from its origin to our plate. In restaurants we want to know who the chef is, that the ingredients are seasonal and sustainably sourced. But I'm often amazed at how invested in understanding the journey of their food someone can be, when in the next sentence they say they buy a supplement from the internet or any old brand from a local health-food store.

I want everyone I work with to understand the role of *everything* that goes into their body – not just food, but supplements too.

Supplement Contamination

This can and does occur in two main ways: through poor quality assurance during production and through deliberate adulteration of otherwise ineffective products.

A landmark study by the International Olympic Committee (IOC)[16] found a high prevalence of contamination in commercially purchased supplements (14.8 per cent).

For professional athletes, contamination means they run the very real risk of inadvertently violating anti-doping rules. An average athlete's career is short; at the very least there could be catastrophic loss of earnings and of course the life-long damage to professional reputation.

OK, you might be thinking, but how does this apply to me?

Well, for starters I'm sure you don't want to consume things that aren't on your supplement's label, any more than you would want ingredients in your food that weren't on the label. And, as well as substances banned for professional athletes, supplements have been found to contain impurities such as lead, broken glass and metal fragments, as well as substances that could potentially cause harm to health, highlighted by the hospital cases above.[17]

How to Minimise Risk

Within professional sport there are policies on engagement with supplements, education for the individual athletes and even clauses about supplement use in their contracts. When working with professional teams and athletes we used to send supplements to labs to do independent batch-testing for contaminants. This obviously isn't an option for everybody, but thankfully there are now quality-assurance schemes, such as **Informed-Sport**.[18] This is an independent organisation that batch-tests products and assesses companies' manufacturing processes, and they make lists of their tested products and brands available to all on their website.

Products they have tested and approved carry the Informed-Sport logo. There are still no guarantees, of course, but as part of a risk-management strategy this should be your first port of call.

Some supplements are more high-risk than others. Those purporting to offer anabolic (muscle-building) results are typically at higher risk of contamination, as are weight-loss supplements and herbal supplements, which lack information about the active ingredients or their purity.[19]

Your Decision-making Process

Any decision around supplementation should weigh up the potential costs (financial, side effects) and benefits (training performance, immunity). Use this process to decide whether supplementation should form part of your Energy Plan.

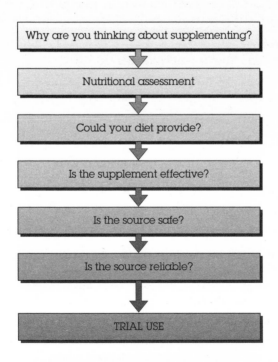

Why are you thinking about supplementing?

Nutritional assessment

Could your diet provide?

Is the supplement effective?

Is the source safe?

Is the source reliable?

TRIAL USE

Adapted from Maughan et al., 2018

Here is a case study to help:

Jonathan is four months into his Energy Plan. His initial goal was to improve his squash game by making himself faster and more mobile around the court. He has organised his performance plates and snacks within his daily and weekly planners and his monitoring indicates that he's reduced his body fat and now consistently has more energy during both training and the working day. For the next goal in his Energy Plan he thinks it might be time to turn to supplementation.

1. THE QUESTION

   Jonathan is now looking to increase his strength and muscle mass, and he thinks creatine may be the answer.

2. NUTRITIONAL ASSESSMENT

   His dietary analysis highlights that he has the key principles in place from following his Energy Plan.

3. COULD THE DIET PROVIDE?

   No. Although he is a meat eater, he can't gain sufficient amounts from his daily intake.

4. IS THE SUPPLEMENT EFFECTIVE?

   Yes. Creatine has strong supporting evidence.

5. IS THE SUPPLEMENT SAFE?

   Yes. The long-term studies around creatine highlight that it is safe when taken following the acceptable protocols.

6. IS THE SOURCE RELIABLE?

   Jonathan went on the Informed-Sport website and sourced a registered product.

   Jonathan made his way through the process and decided to start using the supplement in training.

But what now? When trialling the use of any supplement it's important to **monitor** its effectiveness. In Jonathan's case, it will be important for him to track his weights in the gym, alongside his RPE and body composition. (See the monitoring chapter, page 127, for more information on this.)

And when priorities change, stop taking the supplement. If strength gains are no longer a focus for you, then this decision-making process will point you away from using it. If you aren't run down, there is no need to supplement your vitamin C and zinc all year round. Don't be like the athlete at the start of this chapter, who kept adding unthinkingly to an ever-growing arsenal of supplements. The likelihood is it won't be helping (and may even be hindering). Instead, when your priorities change, go through the process again to reassess your needs.

Now I'd like you to look at your list of any supplements you are currently taking and go through this decision-making process with each one of them. And remember, as with everything to do with your nutrition, if it's not helping your work towards the goals of your Energy Plan – what's it doing there?

CHAPTER 15

Ageing

F or younger readers thinking about skipping this chapter because you feel it doesn't apply to you, remember that the ageing process starts to affect our bodies from our thirties onwards, which is one of the reasons why so many sports professionals retire around that age...

As we reach midlife and our later years, it's time to start thinking about slowing down a bit. Maybe it's OK to give that middle-aged spread a bit of room, to have dessert after all and to have a nice nap on the sofa after dinner rather than getting out for a run or to the local leisure centre for a bit of exercise. Maybe it's time to start thinking seriously about investing in a pair of slippers.

But you try telling that to Ed Whitlock. This 85-year-old crossed the finishing line of the Toronto Marathon in 2016 having run a time of three hours, 56 minutes and 34 seconds – the oldest person ever to run a marathon in less than four hours.

You try telling that to Jack Nicklaus, who won his last major golf championship at the age of 46 in 1986 – and had a really good go at winning another at the age of 58, when he finished sixth at the 1996 Masters.

Try telling that to Martina Navratilova, who in 2006, only a month shy of her fiftieth birthday, won the mixed doubles at the US Open championship – breaking her own record as the oldest ever major tennis champion.

To put Ed Whitlock's time in perspective for the non-marathon runners among us, the average men's time from 2009 to 2014 was four hours, 13 minutes and 23 seconds[1] – and that's for any age group. Amazingly, at the age of 73 he also ran a marathon in under three hours, a time that would have put him in the top 8 per cent of men *of any age* at the 2015 London Marathon. But simply seeing him as an outrageous outlier is not the point. The point is that we can have a hand in shaping how we age – if we adapt our Energy Plan accordingly.

I see this first-hand in my own work. I've recently taken on a client who is attempting her first marathon at the age of 66, and it makes the idea of 'acting your age' seem ludicrous. The idea that you should 'slow down' as you get older is a way of thinking that can be every bit as cancerous as the chronic diseases you're trying to avoid; but it certainly doesn't have to be this way. Your Energy Plan will need to change as your body changes with age, but through adapting it you can equip yourself with the tools to sustain you through midlife and beyond. And the important thing to realise is that **it's never too late to start.**

In mid- and later life you may well be entering the period of your greatest earning potential. A time when you can hopefully start to prioritise the important things in life, like family, that elusive work–life balance, and your health.

In 2018 I was involved in the BBC's Sport Relief, working with four celebrities – presenter Susannah Constantine, comedian Miles Jupp, entertainer Les Dennis and actress Tameka Empson – each of whom had lost their way with their fitness and nutrition. Over 12 weeks we worked with them to get their bodies back.

Each was facing different challenges: Les Dennis was 64 with pre-diabetes, characterised by higher than normal blood glucose. The hormone insulin regulates blood glucose by signalling to the liver and fat cells to absorb levels from the blood. Being overweight, a lack of exercise and poor nutrition (excess calorie intake and carbohydrates, both sugary and starchy) all contribute to the body clearing glucose less efficiently. But with a structured training programme and Energy Plan consisting of more maintenance foods and less fuel, he reduced his body weight by approximately 6 kg (a stone) in 12 weeks, which – importantly – featured a reduction in his visceral fat, the more dangerous fat in the abdominal cavity linked to chronic disease.

Susannah, 56, noted for her fashion prowess, had lost confidence during the menopause, and found her fitness declining and body fat increasing, meaning she couldn't get into the clothes that were part of her brand. Over her 12-week training and Energy Plan, she reduced her body fat by almost a stone (and was able to get back into her favourite dress) and enjoyed increased energy levels and fitness.

For all the contestants the final challenge was a gruelling Tough Guy endurance challenge on an ice-cold January day (I was freezing just watching). They all completed the course that some contestants half their ages were struggling with. In just 12 weeks they had achieved life-changing results.

The challenges I faced with Les and Susannah are representative of a broader problem. Some 77 per cent of men and 63 per cent of women in the 40–60-year-old population in Britain are overweight,[2] and the number of those diagnosed with diabetes has doubled in the last 20 years.[3] As a nation we've become accustomed to thinking – incorrectly – that being overweight is an accepted part of ageing.

So even if you're in your twenties now, it's a good idea to start thinking about addressing some of these issues that you have to come. Or perhaps you can support a loved one to make some meaningful changes to their life. As I hope some of the examples I've used show, you're never too old to start.

Let's take a look at exactly what happens to our bodies as we age.

The Ageing Process

Just as with an ageing car, as the miles begin to add up on our bodies they start to become less efficient. From our thirties onwards, cellular wear and tear results in our tissue and organs beginning to function less effectively. The changes to our bodies include reductions in:

- Aerobic fitness
- Resting metabolism
- Muscle strength and mass (called **sarcopenia**)
- Bone mass
- Vitamin D synthesis (from the skin)
- Thirst sensation

Ageing-related decline in sarcopenia (muscle mass and function), means a decline in your **resting metabolic rate** (the number of calories required to keep your body functioning when at rest). So as each decade passes, you need less fuel to sustain your body every day. How many of us make changes to our nutrition along these lines, though? Without addressing this, and maintaining activity levels, we'll inevitably see an increase in fat mass from the unused fuel (particularly the more dangerous visceral fat around the organs), weight gain and, if we're not careful, a cocktail of chronic diseases including type 2 diabetes and heart disease.

The good news is that we have the means to fight the ageing process. A combination of lifestyle changes can make all the difference, and we are going to focus here on two that are at the heart of the Energy Plan: revving the engine (exercise) and the fuel of the Plan (nutrition). As you get older this is definitely one relationship to invest some time in.

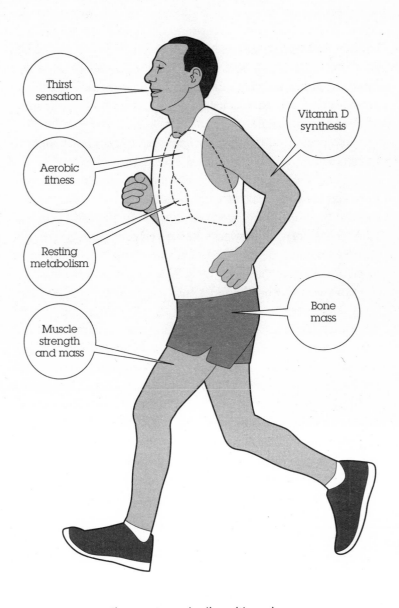

Changes to our bodies with ageing

Lowering Your Energy Consumption

Weight gain as we get older doesn't happen overnight. It comes from running on a positive energy balance for weeks, months and years, and many of us assume it is just a natural part of the ageing process. But by revisiting some of what we talked about in Chapter 1, we can see how managing energy consumption can have a meaningful impact on your weight as you age.

Exercise is a crucial tool in fighting the age-related decline in cells within our organs and tissues. Aerobic exercise (such as jogging, cycling or swimming)[4] improves the heart's function, reducing the risk of cardiovascular disease, and works to keep the brain from conditions involving cognitive decline, such as dementia.[5] Resistance exercise helps to protect muscles from age-related decline and loss. Any type of exercise also increases your metabolism, getting your engine firing. The more active you are, the more fuel you require to meet your daily energy needs. As the stats tell us, physical activity levels are on the decline, with the population now being around 20 per cent less active than in the 1960s,[6] so the first step is to get moving and do something you enjoy – and the more variety the better.

So, although we've covered this before, I'm going to say it again: it's **doubly important** as you age for you to marry your carbohydrate intake to your activity levels for the day. Fuel yourself accordingly: for high-activity days you'll need carbs, but for a day spent at your desk or relaxing with family or friends you won't need them as much. And if you're in your thirties, forties or beyond, you'll need *even less* of them than you would have ten years before.

The focus here is on getting your meals right to regulate blood glucose and insulin, and provide the right fuels for the body, such as protein for maintenance.

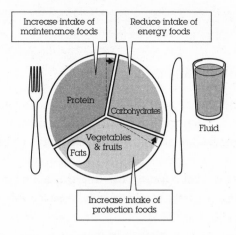

The ageing plate

Reduce the Fuel on Less Intense Days

If you're just hanging out at home on a low-key Saturday, reduce the amount of energy-rich fuel (carbohydrate) you're consuming. The best opportunities to reduce carbs usually come our way at lunch or dinner, as in these meals we can easily compensate by upping our portions of protein and vegetables. As we will discuss further, maintaining protein intake is still crucial on these days, and including protein-rich snacks such as yoghurt, mixed nuts, edamame or even a protein shake is important.

Fuel Up Early for the Hard Days

Instead of letting your evening meal dominate your diet, look at eating more at breakfast and lunch to provide opportunities to fuel the body for harder days. As ever, the type of carbohydrate is vital: focus on adding lower-GI carbs such as boiled sweet potato (lower-GI than baked), rye bread, porridge oats, basmati rice, quinoa (it's definitely never too late to start with this fashionable grain), buckwheat and lentils for a sustained energy release. Reduce higher-GI carbs such as white bread, potato, rice cakes, crackers, bagels, cakes, doughnuts, croissants and most packaged breakfast cereals.

Muscle Maintenance

Like the ageing car we all are as we move through our thirties, we become less efficient, and a key part of this physiological decline is sarcopenia, the loss of muscle mass and function. Sarcopenic muscle loss proceeds at a rate of around 0.8 per cent a year, and strength is lost at a rate of 1–3 per cent a year. This decline is most noticeable from your forties onwards.[7]

In real terms sarcopenia can have a marked impact on your quality of life. This gradual decline in muscle strength and mass might not even be noticeable at first, but just in the same way that weight gain can creep up on you thanks to an energy surplus over several years, you might start to find yourself not quite hitting the heights on the golf course that you once did, or struggling towards the end of your weekend run. And later in life sarcopenia can have serious repercussions, reducing mobility and the ability to perform daily activities you take for granted now, such as climbing the stairs. It also increases progression towards frailty, falls and metabolic disease.[8] All of which might sound a bit doom and gloom for a chapter supposedly about the rich potential of your later years! So let's have a look at what we can do about it.

Resistance Training

Resistance training (or strength training, or weights) is the first step in our plan to get ageing's effects on our muscles under control.

When we make our muscles work against an opposing force, such as a weight, the stress on the muscle cells causes them initially to break down before repairing and growing back stronger. The current UK exercise guidelines[9] are for two sessions a week of resistance training – and these should activate the six major muscle groups: legs, abdominals, back, chest, shoulders and arms. These exercises can easily be completed at home, to get your dose, or even through a class at the local gym or leisure centre, such as body pump.

Often people who have prioritised aerobic training throughout their lives make the mistake of disregarding this kind of training. For example, runners who've always pounded the pavements for serious distances might feel like they're in the shape of their lives as they enter middle age, but if they're not doing any resistance training then they are at risk of sarcopenia too. And it can so easily be addressed with that exercise 'minimum dose' from back in Chapter 4, which can be incorporated into a routine without too much burden.

Do note though that, as we'll come to shortly, that dose might need to increase as we get older.

Maintenance Fuel

Training on its own isn't enough, of course; we need fuel to fight the ageing process too. And it comes in the consumption of **maintenance** foods – protein. When foods containing protein are digested, they are broken down into small building blocks called amino acids, which are then transported in the blood and taken up by the muscle, to be used to repair and build new muscle tissue. We call this process muscle growth – or **muscle protein synthesis** – and we initially looked at it back in Chapter 2.

Your muscles are constantly undergoing a cycle of growth and repair, so your daily intake needs to meet this demand and also maintain your overall muscle mass. The current recommended daily protein intake for sedentary people is 0.75g per kg of body weight; however, recent reviews by leading protein metabolism research groups suggest this is too low for the ageing and elderly population, with their recommendation being a minimum of 1.2 per kg of body weight.[10]

So for someone weighing 70 kg, that's a minimum of 84 g per day, which in practice means an approximately 20 g dose four times a day, for breakfast, lunch, a snack and dinner. As an example, a small chicken breast or a large serving of low-fat Greek yoghurt provides this dose.

The reality of the current situation is that 25 per cent of older men and 50 per cent of older women are still falling short of the RDA[11] – let alone what the body requires to optimise growth following training.

Breakfast and lunch are the meals to target when looking at protein intake, as research shows us that people find it easier to eat enough in their evening meal, while meals earlier in the day fall short – particularly breakfast (a breakfast of toast and jam, for example, would not provide enough protein).[12] Adding a serving of low-fat Greek yoghurt or a whey protein shake (these aren't just for athletes) can easily increase your intake.

So which are the best foods for muscle maintenance? As we saw in Chapter 1, complete proteins – those that contain a complete range of amino acids – have been shown to be the most beneficial. This means foods such as dairy, poultry and fish, as well as quinoa and buckwheat. Other plant sources such as rice and beans are good sources but are missing at least one amino acid, which means they need to be combined to make them complete. See 'Performance Plates', page 79.

This combination of resistance training and sufficient protein is exactly the same principle that elite athletes follow to keep their muscles strong and powerful. It is worth noting here that ageing causes the muscles to become less responsive (more set in their ways) to both training and protein. This is known as 'anabolic resistance'[13] and it means a bigger dose of either training or protein intake may be required to maintain the body. The best first step is to ensure a good source of protein with each meal and snack.

BUT WHAT TYPE OF PROTEIN?

Working out which type of protein foods to focus on can be confusing. See the protein guide in the appendix for the full breakdown but, in general, it's good to include more fish, especially oily fish containing omega 3 fatty acids, such as salmon, tuna (fresh, not canned), mackerel, sardines, trout and herring. Although the evidence is still emerging, the studies so far suggest that they also reduce inflammation and potentially lower the risk of heart disease, cancer and arthritis, all very real risks as we get older.

Replacing red meat (especially processed meat such as bacon, sausages, ham and salami) is recommended due to the increased risk of bowel cancer linked to these types of meat. Chicken, fish and fresh, non-processed meat are better. And a combination of beans, tofu, lentils, eggs and higher-protein grains, such as quinoa and buckwheat, can all increase the protein content of your meal whether you're vegetarian, vegan or just want to cut down your meat consumption.

Protection Foods

So far we've looked at **reducing your energy foods** (carbs – particularly the simple sugars) and **increasing your maintenance foods** (protein). And now we need to address the third component of your fuel, the **protection** foods (micronutrients).

When we're young, it's easy to think of things like heart disease, diabetes and cancer as almost abstract concepts – things that happen to much older, *other* people. But as we age these conditions sadly become more a fact of life, for friends and family members and even for ourselves. And it's as we enter later life that the pay-off for including more fruit and vegetables in our diet becomes clearer.

Five a Day (Or More)

Increased fruit and vegetable intake is linked to a reduction in these kinds of conditions. And the pay-off is potentially huge. A large public health study looked at the eating habits of 65,226 people and found a significant relationship between eating fruit and vegetables and death at any age. Eating five to seven portions a day reduced risk of death by 36 per cent, and for those hitting more than seven portions a day, risk of death was reduced by 42 per cent.[14] And more recently it's been found that the benefits are higher again at ten portions a day.[15]

But let's get real here for a second and put this into perspective. The latest UK findings, published in March 2017, show that only 26 per cent of adults are reaching five or more portions of fruit and veg each day.[16] And with there being benefits for each extra portion of fruit and veg we consume each day, we need to shift the focus to increasing our intake, not aiming for an unrealistic amount we'll never get through. We're looking for progress, not perfection.

HOW CAN I INCREASE MY FRUIT AND VEGETABLE INTAKE?

1. **On the list:** Make sure that you always have a supply of your favourite vegetables and fruit available to add at mealtimes.
2. **Function first:** Understand the role of specific protection foods - from reducing muscle soreness

---

to boosting your immunity – rather than simply thinking of them as being 'healthy', to give you sound reasoning to include more in your diet.

3. **Double up:** With our performance plates, we recommend two servings of two different vegetables with lunch and dinner. If you're able to hit this, along with a serving of fruit at breakfast, you'll hit your five-a-day.

4. **Progress not perfection:** It may be that just including one portion with each meal is a big first step for you, and that's fine. Remember, every extra portion yields health benefits, so this is a great start!

5. **Get creative:** Trying out new recipes is great for boosting your fruit and vegetable intake, and also an effective way of adding new things that haven't been on your shopping list before.

6. **Breakfast toppers:** Antioxidant-rich berries, such as blueberries, raspberries and strawberries, are an easy topping at breakfast time or as part of a snack. Keep your fridge stocked, or buy frozen as a cheaper alternative.

But why are fruit and vegetables so good for us? Well, for starters they provide the fibre to maintain a healthy digestive tract and prevent constipation, and they contain a host of vitamins and minerals to support the health of all the body's tissues. But more recently the importance of the phytochemicals (or phytonutrients, if you prefer) found in fruit and vegetables (phyto refers to the Greek word for plants) has been discovered. Most research so far has looked at the polyphenol family (which includes flavonoids, anthocyanins and others) and another group called carotenoids, which includes lycopene (which tomatoes are rich in) and beta-carotene.

WHICH SHOULD YOU BE REACHING FOR?

Essentially each colour of vegetable has different benefits:

- **Root vegetables** (e.g. carrots, beetroots, sweet potato) are **high in fibre** for a healthy digestive system, keeping food passing through and reducing constipation.
- **Cruciferous vegetables** ('your greens' – such as broccoli, kale, rocket, bok choi, cabbage) are high in **glucosinolates,** which activate enzymes that may protect against cancer.
- **Red and purple fruits and vegetables** (e.g. berries) – are high in **antioxidants** and also **anthocyanins**, which dilate blood vessels, improve cardiovascular health and may reduce DNA damage. **Tomatoes** are high in the antioxidant **lycopene**, which can reduce LDL ('bad') cholesterol and improve heart health. Red and black **grapes** deliver **resveratrol**, which has anti-cancer properties, and **peppers and chillies** are great vitamin C providers.
- The oranges, greens and yellows of **citrus fruits** are high in **vitamin C**, which boosts immunity and collagen production for strong connective tissue including ligaments and tendons.
- **Lighter coloured fruit and vegetables** (such as onions, grapes and apples) and also **green tea** all contain **quercetin,** which has been shown to have beneficial effects on immunity.

We've already discussed the idea of 'eating the rainbow', and that kind of variety is key here to ensure a full range of **protection** benefits. Ensure that you are including at least three colours in each meal to optimise your intake.

Feeling it in Your Bones

It's not just our muscles that feel the effects of ageing. Bone density peaks around our mid-thirties, and from then on there is a gradual decline; the rates of breakdown begin to exceed growth. **Osteopenia**, a condition in which your bones are weaker than normal, is most likely to occur in your fifties; it shouldn't be confused with **osteoporosis**, a more serious condition where your bones are significantly weaker and more likely to fracture. Although women lose bone density rapidly in the first few years after the menopause, and are at greater risk of osteoporosis, men can also suffer from it. Nutrition, particularly in the form of micronutrients, and exercise can both play an important part in keeping bones strong.

With your exercise, keep it weight-bearing. Resistance training counts as weight-bearing, as do things like tennis, running and dancing.

And with nutrition, the micronutrients calcium and vitamin D have a big part to play. Calcium is, of course, famous for its role in bone health – 99 per cent of our body's calcium is found in our bones. The RDA for adults is 700 mg (achievable by consuming three glasses of milk or calcium-enriched milk alternatives like soya, oat or nut). This RDA increases to up to 1,200 mg per day in menopausal women and men over 55.[17]

There are other sources of calcium than milk, of course – dairy such as yoghurt and cheese, tofu, fortified soya, rice and oat drinks, and vegetables such as broccoli and spring greens are all good sources.

But consuming calcium alone isn't the answer. It needs its partner, vitamin D, so that our bodies can effectively absorb the calcium and build new bone tissue. Dubbed the 'sunshine vitamin' because it is produced by exposure to sunlight, vitamin D is a true protection nutrient, improving immunity and muscle function. There is also emerging evidence to show its impact on chronic conditions such as cardiovascular disease and type 2 diabetes, as well as conditions of cognitive decline such as dementia and Alzheimer's disease.

Some bad news, again, however: ageing reduces the ability of our skin to synthesise vitamin D, meaning that deficiencies can occur. Your vitamin D level is easily discovered by requesting a blood test from your doctor, and the table of micronutrients in Chapter 2 gives more information on foods containing vitamin D and supplementation.

Liquid Assets

Part of the host of physiological changes that occur with ageing is that total body water reduces.[18] Also, as we get older our perception of thirst reduces, and the kidneys' ability to maintain body water is diminished. All these things together mean that staying hydrated can become more challenging in later years, and if you exercise regularly, drinking only when you feel thirsty may not be enough to offset dehydration.

For more information on developing your personalised hydration plan, head to page 127 (Moving the Needle: Monitoring Your Progress chapter).

ENERGY EXPRESS: YOUR AGEING PLAN

- **Rev the engine each day.** On days where you don't have time for a fitness class or a run, just increasing your step count (for example, with a brisk 30-minute walk at lunchtime) can make a big difference to reducing body fat.
- **Don't resist resistance.** Aim for two resistance sessions a week, at the gym or at home, to activate the six major muscle groups: legs, abdominals, back, chest, shoulders and arms. Book

a session with a personal trainer to take you through the exercises to build confidence, or try a class such as body pump.

- **Less exercise = less carbs.** If you've had a day stuck behind your desk or relaxing with family and friends, cut down on the fuel. Often the easiest way is to reduce carbs at dinner. Just increase your portion of protein (fish, poultry, pulses, tofu) and mixed vegetables to bulk out the meal. And cut out the snacking between meals.
- **Prioritise the protein.** There are lots of ways to achieve this: a portion of lean meat (or fish or poultry); including extra nuts, seeds or beans in your meal; some yoghurt or eggs with breakfast.
- **Up the protection.** Protection foods can support your immune system, muscles and bones on a daily basis as well as warding off chronic diseases. Aim for at least two varieties (colours) of vegetables with each meal to maximise the protective benefits.
- **Pace yourself.** The points above are the big hitters around training and nutrition, so bring them in over time. For example, it is natural for building resistance training into your routine to take a while – it's a big step.

Conclusion:
Evolving Energy

Coming to the end of this book might feel like the end of this part of the journey, but in reality it's just the beginning.

With performance nutrition, as with any branch of sports science, no matter how much experience you acquire, nothing ever stands still. There are always new scientific studies, new sources of knowledge to apply and make meaningful changes to performance. I think it's fair to say that I'm still learning and evolving, and always will be.

Your Energy Plan will not be standing still either. As time moves on and your lifestyle, work and needs change, and your body itself changes, so your Plan will need to evolve. It might mean that eventually your goal becomes to have enough energy to meet the demands of family life; it could equally mean getting into a new sport such as swimming or cycling, and wanting to tailor your Plan to yield maximum results from that. It might be that you change career or your work becomes more flexible, and you look at ways to adapt your Energy Plan on different days to accommodate your new lifestyle.

There will of course be lots of new scientific research that will be seized upon by a diet and lifestyle industry desperate to continually keep you informed about the new buzz superfoods

and the latest way to lose weight. Some of these new developments will be of use, others less so, but I hope that through
reading this book you now have a better sense of how to inter-
rogate these stories and get a sense of what might work for you
and what won't, as part of a balanced Energy Plan.

Because while all sorts of new developments in nutrition lie
in front of us, the fundamentals of the Energy Plan will never
go out of date. Using your food as fuel to meet the demands
of your day and week; building your meals around protein
and adjusting your carbohydrate intake accordingly; eating
a broad range of micronutrients in your vegetables and fruit;
and sticking to the mono- and polyunsaturated sources of
fats in your meals. Being flexible in your approach, so that
as your needs change, your Plan can too. Relaxing your Plan
from time to time, just like the elite athletes I work with do
when they go on holiday at the end of a season or after a big
event. As I've said throughout this book, and I'll say for the
last time now, living well doesn't mean you can't enjoy some
of the other things life has to offer.

These are the cornerstones of a sustainable and effective
approach to nutrition and lifestyle that will last you throughout your life, whichever corner of the globe you visit or live
in. Through adopting your own Energy Plan, you are going to
reap the rewards of improved energy, the sustained peaks and
reduced troughs that will enable you to be the best version of
yourself in your work and play, every day.

You now have the skills to make every meal a performance
plate, to build these meals and snacks into your days and
weeks to feel energised and able to meet any challenge. You
know how to use the performance-enhancing foods, drinks
and supplements at your disposal, and how to decide if and
when you might need them. You have the skills to land at a
destination anywhere in the world and deliver a great performance; to keep your immune system robust so that you can
feel well all year round; and to do the one thing well that
affects our energy more than almost anything else, and which

we often take for granted: sleep. Most importantly, you are now able to monitor all of this and see when you need to make changes, and when you are making progress. You have the skills to manage this on your own now.

So, no matter where you are in your journey, whether you've just cleared the cupboards and are preparing to start in earnest or you're now well into this process of feeling the improved changes to energy, mood and health that I assure you this book offers, I wish you the very best of luck with your Energy Plan.

Appendix

Carbohydrates

CARBOHYDRATE SOURCE	APPROXIMATE WEIGHT	GI RANKING
CEREALS _Each providing 30 g carbohydrate_		
Oats	50 g	Low
Muesli, no added sugar	50 g	Low
All Bran	60 g	Low
Bran Flakes	45 g	Medium
Weetabix	2 biscuits	Medium
Shredded Wheat	2 biscuits	High
Cornflakes	35 g	High
Granola	40 g	High
BREADS _each providing 30 g carbohydrate_		
Sourdough rye	2 slices	Low
Granary, medium sliced	2 slices	Low

All weights listed are uncooked.

Pitta	1 medium sized	Medium
White bread, medium sliced	2 slices	High
Bagel	55 g	High
English muffin	1	High
French baguette	55 g	High
GRAINS & PULSES *Each providing 30 g carbohydrate*		
Rice, wild	45 g	Low
Bulgur	45 g	Low
Buckwheat	40 g	Low
Quinoa	60 g	Low
Lentils	60 g	Low
Barley	40 g	Low
Farro	50 g	Low
Spelt	45 g	Low
Couscous	40g	Medium
Rice, basmati	40 g	Medium
Rice, white, instant	40 g	High
PASTA, NOODLES & POTATOES *Each providing 30 g carbohydrate*		
Spaghetti, wholewheat	50 g	Low
Fusilli, penne, linguine, wholewheat	50 g	Low
Noodles	40 g	Medium
Gnocchi	90 g	Medium
White potato, boiled	150 g	High
Sweet potato, boiled	150 g	Low

FRUITS		
Each providing 10 g carbohydrate		
Blueberries	100 g	Low
Strawberries	170 g	Low
Raspberries	200 g	Low
Apple	80 g	Low
Orange	120 g	Low
Pear	80 g	Low
Kiwi	90 g	Low
Half banana (small)	100 g	Medium
Pineapple	100 g	Medium
Raisins	15 g	Medium
Melon	180 g	High
Dates	15 g	High
VEGETABLES *Most vegetables are low in carbohydrate and are low-GI. An 80g portion of non-starchy vegetables <10 g of carbohydrate.*		Low

Animal Protein

PROTEIN SOURCE *Each provides 20 g protein*	APPROXIMATE WEIGHT
Skinless, chicken breast, small	85 g
Turkey breast, small	80 g
Turkey breast mince (2% fat)	80 g
Ham, lean	110 g
Beef lean steak mince (5% fat)	85 g
Steak, lean	90 g
Pork, lean	85 g
Lamb, lean	75 g
Salmon, fresh	95 g
Salmon, canned	100 g
White fish, baked	100 g
Tuna steak	85 g
Tuna, canned in spring water	85 g
King prawns	100 g
Eggs, whole, free-range (3)	160 g
Cottage cheese, fat-free	200 g
Cows' milk (skimmed, semi-skimmed, whole)	568 ml (pint)
Low-fat Greek yoghurt	200 g

Plant Protein

Combine two options in the below list in each meal. Options in italics are incomplete proteins and need to be combined.

PROTEIN SOURCE Each example provides 10 g protein	APPROXIMATE WEIGHT
Tofu	120 g
Tempeh	50 g
Quinoa	70 g
Buckwheat	75 g
Edamame	100 g
Kidney beans (canned)	*120 g*
Black beans (canned)	*110 g*
Chickpeas (canned)	*120 g*
Lentils (canned)	*110 g*

Fats

FATS SOURCE *Each example provides 10 g fat*	APPROXIMATE WEIGHT	HANDY MEASURE
Extra-virgin olive oil	10 ml	$^2/_3$ tablespoon
Seeds (chia, flax, sunflower)	20 g	1 heaped tablespoon
Mixed nuts (almond, macadamia, pistachio, walnut)	20 g	8–10 nuts
Avocado	50 g	½ small-sized
Nut butter	20 g	1 tablespoon

Acknowledgments

To Sam Jackson, Joel Rickett and all at Penguin Random House for supporting my direction and recognising the need for this book.

To Steve Burdett for his words, collaboration and also his patience. I know we took the hard road to get the detail right on this.

Thank you to Richard Pike from C&W for his support in developing this vision, and his guidance throughout the process. To Leah Feltham for her help with editing and Dr Scott Robinson for early proofreading.

To my family whose ongoing support makes everything possible: Christine, Andy, Kirsty, Dave, Finlay and Soraya (I battled hard not to use nicknames here!).

To all of my friends, in London and beyond – and to those who shaped my early years at The Castle School and Richard Huish College. And to Professor Clyde Williams OBE at Loughborough University for showing me that a career in sports nutrition was for me.

Finally to my trailblazing colleagues and friends who challenge and push the boundaries – Arsène Wenger OBE, Professor Greg Whyte OBE, Dr Alan McCall, Dr Ben Rosenblatt, Ben Ashworth, David Priestley and Huss Fahmy.

A final thank you to all of my clients and colleagues down the years, who believed nutrition can make a difference – you have contributed to this book.

References

Part I: The Energy Balance

Chapter 1. The Engine

1. de Ataide e Silva T, Di Cavalcanti Alves de Souza ME, de Amorim JF, Stathis CG, Leandro CG and Lima-Silva AE; 'Can carbohydrate mouth rinse improve performance during exercise? A systematic review'; *Nutrients*, Jan 2014

Chapter 2. The Fuels

1. Magistretti PJ and Allaman I; 'A Cellular perspective on brain energy metabolism and functional imaging'; *Neuron Review*, May 2015
2. Scientific Advisory Committee on Nutrition; 'Draft report: Saturated fats and health'; Public Health England, July 2018
3. Zong G, Li Y, Sampson L, Dougherty LW, Willett WC, Wanders AJ, Alssema M, Zock PL, Hu FB and Sun Q; 'Monounsaturated fats from plant and animal sources in relation to risk of coronary heart disease among US men and women'; *Am J Clin Nutr.*, Mar 2018
4. ibid.

5. Berger ME, Smesny S, Kim S-W, Davey CG, Rice S, Sarnyai Z, Schlögelhofer M, Schäfer MR, Berk M, McGorry PD and Amminger GP; 'Omega-6 to omega-3 polyunsaturated fatty acid ratio and subsequent mood disorders in young people with at-risk mental states: a 7-year longitudinal study'; *Translational Psychiatry, Aug* 2017

6. Calder PC; 'Omega-3 fatty acids and inflammatory processes'; *Nutrients,* Mar 2010

7. Tachtsis B, Camera D and Lacham-Kaplan O; 'Potential roles of n-3 PUFAs during skeletal muscle growth and regeneration'; *Nutrients,* Mar 2018

8. Casal S, Malheiro R, Sendas A, Oliveira BP and Pereira JA; 'Olive oil stability under deep-frying conditions'; *Food Chem Toxicol.,* Oct 2010

9. Thomas DT, Erdman KA and Burke LM; 'Position of the Academy of Nutrition and Dietetics, Dietitians of Canada, and the American College of Sports Medicine: Nutrition and Athletic Performance'; *J Acad Nutr Diet.,* Mar 2016

10. ibid.

11. Stokes T, Hector AJ, Morton RW, McGlory C and Phillips SM; 'Recent perspectives regarding the role of dietary protein for the promotion of muscle hypertrophy with resistance exercise training'; *Nutrients,* Feb 2018

12. Areta JL, Burke LM, Ross ML, Camera DL, West DW, Broad EM, Jeacocke NA, Moore DR, Stellingwerff T, Phillips SM, Hawley JA and Coffey VG; 'Timing and distribution of protein ingestion during prolonged recovery from resistance exercise alters myofibrillar protein synthesis'; *J Physiol.,* May 2013

13. Thomas DT, Erdman KA and Burke LM; 'American College of Sports Medicine Joint Position Statement. Nutrition and Athletic Performance'; *Med Sci Sports Exerc.,* Mar 2016

14. Sawka MN and Montain SJ; 'Fluid and electrolyte supplementation for exercise heat stress'; *Am. J Clin. Nutr., Aug* 2000

Chapter 3. The Accelerator

1. Burke LM, Ross ML, Garvican-Lewis LA, Welvaert M, Heikura IA, Forbes SG, Mirtschin JG, Cato LE, Strobel N, Sharma AP and Hawley JA; 'Low carbohydrate, high fat diet impairs exercise economy and negates the performance benefit from intensified training in elite race walkers'; *J Physiol.*, May 2017
2. Ivy JL, Katz AL, Cutler CL, Sherman WM and Coyle EF; 'Muscle glycogen synthesis after exercise: effect of time of carbohydrate ingestion'; *Journal of Applied Physiology.*, Apr 1988
3. Garber CE, Blissmer B, Deschenes MR, Franklin BA, Lamonte MJ, Lee IM, Nieman D and Swain DP; 'Quantity and quality of exercise for developing and maintaining cardiorespiratory, musculoskeletal, and neuromotor fitness in apparently healthy adults: Guidance for prescribing exercise'; *Med Sci Sports Exerc.*, Jul 2011
4. Meeusen R, Duclos M, Foster C, Fry A, Gleeson M, Nieman D, Raglin J, Rietjens G, Steinacker J and Urhausen A; European College of Sport Science; American College of Sports Medicine; 'Prevention, diagnosis, and treatment of the overtraining syndrome: joint consensus statement of the European College of Sport Science and the American College of Sports Medicine'; *Med Sci Sports Exerc.*, Jan 2013
5. Mountjoy M, Sundgot-Borgen J, Burke L, Carter S, Constantini N, Lebrun C, Meyer N, Sherman R, Steffen K, Budgett R and Ljungqvist A; 'The IOC consensus statement: beyond the female athlete triad–relative energy deficiency in Sport (RED-S)'; *Br J Sports Med.*, Apr 2014

Part II: Your Energy Plan

Chapter 4. Getting Started

1. Schoenfeld BJ, Contreras B, Krieger J, Grgic J, Delcastillo K, Belliard R and Alto A; 'Resistance training volume enhances muscle hypertrophy'; *Med Sci Sports Exerc.*, Aug 2018

Chapter 5. Performance Plates: Different Fuels for Different Days

1. Maughan RJ, Burke LM, Dvorak J, Larson-Meyer DE, Peeling P, Phillips SM, Rawson ES, Walsh NP, Garthe I, Geyer H, Meeusen R, van Loon LJC, Shirreffs SM, Spriet LL, Stuart M, Vernec A, Currell K, Ali VM, Budgett RG, Ljungqvist A, Mountjoy M, Pitsiladis YP, Soligard T, Erdener U and Engebretsen L; 'IOC consensus statement: dietary supplements and the high-performance athlete'; *BJSM,* Apr 2018
2. ibid.
3. Duncan MJ, Lyons M and Hankey J; 'Placebo effects of caffeine on short-term resistance exercise to failure'; *Int J Sports Physiol Perform.*, Jun 2009
4. Kamimori GH, Karyekar CS, Otterstetter R, Cox DS, Balkin TJ, Belenky GL and Eddington ND; 'The rate of absorption and relative bioavailability of caffeine administered in chewing gum versus capsules to normal healthy volunteers'; *Int J Pharm.*, Mar 2002
5. Maughan RJ, Watson P, Cordery PAA, Walsh NP, Oliver SJ, Dolci A, Rodriguez-Sanchez N and Galloway SDR; 'Sucrose and sodium but not caffeine content influence the retention of beverages in humans under euhydrated conditions'; *Int J Sport Nutr Exerc Metab.*, Oct 2018 *and* Killer SC, Blannin AK and Jeukendrup AE; 'No evidence of dehydration with moderate daily coffee intake: A counterbalanced cross-over study in a free-living population'; *Plos One.*, 2014 Jan

6. Starbucks, Beverage Nutrition Information, Autumn 2018; www.starbucks.co.uk/quick-links/nutrition-info

7. Ganio MS, Klau JF, Casa DJ, Armstrong LE and Maresh CM; 'Effect of caffeine on sport-specific endurance performance: A systematic review'; *Journal of Strength and Conditioning Research,* Jan 2009

8. Spriet LL; 'Exercise and sport performance with low doses of caffeine'; *Sports Medicine,* Nov 2014

9. Vital-Lopez FG, Ramakrishnan S, Doty TJ, Balkin TJ and Reifman J; 'Caffeine dosing strategies to optimize alertness during sleep loss'; *J Sleep Res.,* Oct 2018

10. 2b-alert-web.bhsai.org/2b-alert-web/login.xhtml

11. Parr EB, Camera DM, Areta JL, Burke LM, Phillips SM, Hawley JA and Coffey VG; 'Alcohol ingestion impairs maximal post-exercise rates of myofibrillar protein synthesis following a single bout of concurrent training'; *Plos One,* Feb 2014

Chapter 7. Winning Behaviours

1. Borek AJ, Abraham C, Greaves CJ and Tarrant M; 'Group-based diet and physical activity weight-loss interventions: A systematic review and meta-analysis of randomised controlled trials'; *IAAP.,* Feb 2018

2. Braude L and Stevenson RJ; 'Watching television while eating increases energy intake. Examining the mechanisms in female participants'; *Appetite,* May 2014

3. Spence C; *Gastrophysics: The New Science of Eating*; Penguin, Mar 2017

4. Holden S, Zlatevska N and Dubelaar C; 'Whether smaller plates reduce consumption depends on who's serving and who's looking: A meta-analysis'; *Journal of the Association for Consumer Research,* Jan 2016

5. Michel C, Velasco C and Spence C; 'Cutlery matters: heavy cutlery enhances diners' enjoyment of the food served in a realistic dining environment'; *Flavour,* Jul 2015

Chapter 8. Moving the Needle: Monitoring Your Progress

1. Gidde L, Leidner D and Gonzalez E; 'The role of fitbits in corporate wellness programs: Does step count matter?'; Hawaii International Conference on System Sciences, Jan 2017
2. National Institute for Health and Care Excellence; 'Obesity: Identification, assessment and management of overweight and obesity in children, young people and adults', Nov 2014
3. Saw AE, Main LC and Gastin PB; 'Monitoring the athlete training response: subjective self-reported measures trump commonly used objective measures: A systematic review'; *Brit J Sports Med.*, Mar 2016
4. Fokkema T, Kooiman TJ, Krijnen WP, Van der Schans CP and DE Groot M; 'Reliability and validity of ten consumer activity trackers depend on walking speed', *Med Sci Sports Exerc.*, Apr 2017
5. Pettitt C, Liu J, Kwasnicki RM, Yang GZ, Preston T, and Frost GS; 'A pilot study to determine whether using a lightweight, wearable micro-camera improves dietary assessment accuracy and offers information on macronutrients and eating rate', *Br J Nutr.*, 2016

Chapter 9. On-Plan in the Workplace

1. Office for National Statistics; 'Trends in self-employment in the UK: Analysing the characteristics, income and wealth of the self-employed', Feb 2018
2. De Castro JM; 'When, how much and what foods are eaten are related to total daily food intake'; *Br J Nutr.*, Oct 2009
3. Bo S, Musso G, Beccuti G, Fadda M, Fedele D, Gambino R, Gentile L, Durazzo M, Ghigo E and Cassader M; 'Consuming more of daily caloric intake at dinner predisposes to obesity. A 6-year population-based prospective cohort study'; *Plos One,* Sep 2014

4. Anderson L, Orme P, Naughton, RJ, Close GL, Milsom J, Rydings D, O'Boyle A, Di Michele R, Louis J, Morgans R, Drust B and Morton JP; 'Energy intake and expenditure of professional football players of the English Premier League: evidence of carbohydrate periodization'; *Int J Sport Nutr Exerc Metab.*, 2017

5. Ocado Group; 'A nation of aspiring foodies stick in a nine-meal rut'; Feb 2015

6. Haghighatdoost F, Azadbakht L, Keshteli AH, Feinle-Bisset C, Daghaghzadeh H, Afshar H, Feizi A, Esmaillzadeh A and Adibi P; 'Glycemic index, glycemic load, and common psychological disorders'; *Am J Clin Nutr.*, Jan 2016

7. Penckofer S, Quinn L, Byrn M, Ferrans C, Miller M and Strange P; 'Does glycemic variability impact mood and quality of life?'; *Diabetes Technol Ther.*, Apr 2012

Chapter 10. From Plan to Plate

1. www.bbcgoodfood.com/howto/guide/marathon-meal-plans

2. www.jamescollinsnutrition.com/recipes

3. Soil Association; 'The organic market report'; 2018

4. Organic Trade Association survey; May 2018

5. Dangour AD, Dodhia SK, Hayter A, Allen E, Lock K and Uauy R; 'Nutritional quality of organic foods: a systematic review'; *The American Journal of Clinical Nutrition,* Sep 2009

Part III: Sustainable Energy

Chapter 11. Recharging

1. Cooke R; '"Sleep should be prescribed": What those late nights out could be costing you'; *Observer,* Sep 2017

2. Hirshkowitz M, Whiton K, Albert SM, Alessi C, Bruni O, DonCarlos L, Hazen N, Herman J, Adams Hillard PJ, Katz ES, Kheirandish-Gozal L, Neubauer DN, O'Donnell

AE, Ohayon M, Peever J, Rawding R, Sachdeva RC, Setters B, Vitiello MV and Ware JC; 'National Sleep Foundation's updated sleep duration recommendations: final report'; *Sleep Health*, Dec 2015

3. Halson SL; 'Sleep in elite athletes and nutritional interventions to enhance sleep'; *Sports Med.*, 2014

4. Walker MP and Stickgold R; 'It's practice, with sleep, that makes perfect: Implications of sleep-dependent learning and plasticity for skill performance'; *Clin. Sports Med.*, 2005

5. Fullagar HH, Skorski S, Duffield R, Julian R, Bartlett J and Meyer T; 'Impaired sleep and recovery after night matches in elite football players'; *J Sports Sci.*, Jul 2016

6. Fietze I, Strauch J, Holzhausen M, Glos M, Theobald C, Lehnkering H and Penzel T; 'Sleep quality in professional ballet dancers'; *Chronobiol Int.*, Aug 2009 *and* Hausswirth C, Louis J, Aubry A, Bonnet G, Duffield R and LE Meur Y; 'Evidence of disturbed sleep and increased illness in over-reached endurance athletes'; *Med Sci Sports Exerc.*, 2014

7. Taheri S, Lin L, Austin D, Young T and Mignot E; 'Short sleep duration is associated with reduced leptin, elevated ghrelin, and increased body mass index'; *Plos Med.*, Dec 2004

8. Benedict C, Hallschmid M, Lassen A, Mahnke C, Schultes B, Schiöth HB, Born J and Lange T; 'Acute sleep deprivation reduces energy expenditure in healthy men'; *Am J Clin Nutr.*, Jun 2011

9. Halson SL; 'Sleep in elite athletes and nutritional interventions to enhance sleep'; *Sports Med.*, 2014

10. ibid.

11. Howatson G, Bell PG, Tallent J, Middleton B, McHugh MP and Ellis J; 'Effect of tart cherry juice (Prunus cerasus) on melatonin levels and enhanced sleep quality'; *Eur J Nutr.*, Dec 2012

12. Pietilä J, Helander E, Korhonen I, Myllymäki T, Kujala UM and Lindholm H; 'Acute effect of alcohol intake on cardiovascular autonomic regulation during the first hours

of sleep in a large real-world sample of Finnish employees: Observational study'; *JMIR Ment. Health,* 2018

13. Res PT, Groen B, Pennings B, Beelen M, Wallis GA, Gijsen AP, Senden JM and Van Loon LJ; 'Protein ingestion before sleep improves postexercise overnight recovery'; *Med Sci Sports Exerc.,* Aug 2012

14. Reyner LA and Horne JA; 'Suppression of sleepiness in drivers: combination of caffeine with a short nap'; *Psychophysiology,* Nov 1997

15. Cheri D. Mah, Kenneth E. Mah, Eric J. Kezirian and William C. Dement; 'The effects of sleep extension on the athletic performance of collegiate basketball players'; *Sleep,* 2011 Jul

16. Åkerstedt T, Ghilotti F, Grotta A, Zhao H, Adami HO, Trolle-Lagerros Y and Bellocco R; 'Sleep duration and mortality: Does weekend sleep matter?'; *Journal of Sleep Research,* May 2018

17. Wittmann M, Dinich J, Merrow M and Roenneberg T; 'Social jetlag: misalignment of biological and social time'; *Chronobiol Int.,* 2006

Chapter 12. Immunity

1. Gleeson M; 'Immunological aspects of sport nutrition'; *Immunol Cell Biol.,* Feb 2016

2. Walsh NP; 'Recommendations to maintain immune health in athletes'; *Eur J Sport Sci.,* Jul 2018

3. Nedelec, M, McCall, A, Carling, C, Legall, F, Berthoin, S and Dupont, G; 'The influence of soccer playing actions on the recovery kinetics after a soccer match'; *J Strength Cond Res.,* 2014

4. Hawley JA and Burke LM; 'Carbohydrate availability and training adaptation: effects on cell metabolism'; *Exerc Sport Sci Rev.,* 2010

5. Fortes MB, Diment BC, Di Felice U and Walsh NP; 'Dehydration decreases saliva antimicrobial proteins

important for mucosal immunity'; *Applied Physiology Nutrition and Metabolism,* Jun 2012

6. Witard OC, Turner JE, Jackman SR, Kies AK, Jeukendrup AE, Bosch JA and Tipton KD; 'High dietary protein restores overreaching induced impairments in leukocyte trafficking and reduces the incidence of upper respiratory tract infection in elite cyclists'; *Brain Behav Immun.,* Jul 2014

7. Gleeson M; 'Immunological aspects of sport nutrition'; *Immunol Cell Biol.,* Feb 2016

8. Singh M and Das RR; 'Zinc for the common cold'; *Cochrane Database Syst Rev.,* 2013

9. Hemilä H and Chalker E; 'Vitamin C for preventing and treating the common cold'; *Cochrane Systematic Review,* Jan 2013

10. ibid.

11. Scherr J, Nieman DC, Schuster T, Habermann J, Rank M, Braun S, Pressler A, Wolfarth B and Halle M; 'Nonalcoholic beer reduces inflammation and incidence of respiratory tract illness'; *Med Sci Sports Exerc.,* Jan 2012

12. Nieman DC, Henson DA, Gross SJ, Jenkins DP, Davis JM, Murphy EA, Carmichael MD, Dumke CL, Utter AC, McAnulty SR, McAnulty LS and Mayer EP; 'Quercetin reduces illness but not immune perturbations after intensive exercise'; *Med Sci Sports Exerc.,* 2007 Sep

13. White WS, Zhou Y, Crane A, Dixon P, Quadt F and Flendrig LM; 'Modeling the dose effects of soybean oil in salad dressing on carotenoid and fat-soluble vitamin bio-availability in salad vegetables'; *Am J Clin Nutr.,* 2017

14. Hao Q, Dong BR and Wu T; 'Probiotics for preventing acute upper-respiratory tract infections'; *Cochrane Systematic Review,* Feb 2015

15. Cohen S, Tyrrell DA and Smith AP; 'Psychological stress and susceptibility to the common cold'; *N Engl J Med.,* 1991

16. Aric A. Prather, Denise Janicki-Deverts, Martica H. Hall and Sheldon Cohen; 'Behaviorally assessed sleep and susceptibility to the common cold'; *Sleep,* Sep 2015

Chapter 13. Travel

1. Wilson BJ; 'How one man earned over 19 million miles on a single airline'; *Business Insider,* Mar 2018
2. Reid P; 'Catch him if you can – Paul Dunne clocks up the air miles'; *The Irish Times,* Jan 2018
3. Bowden Davies KA, Sprung VS, Norman JA, Thompson A, Mitchell KL, Halford JCG, Harrold JA, Wilding JPH, Kemp GJ8 and Cuthbertson DJ; 'Short-term decreased physical activity with increased sedentary behaviour causes metabolic derangements and altered body composition: effects in individuals with and without a first-degree relative with Type 2 diabetes'; *Diabetologia,* Jun 2018
4. McGlory C, von Allmen MT, Stokes T, Morton RW, Hector AJ, Lago BA, Raphenya AR, Smith BK, McArthur AG, Steinberg GR, Baker SK and Phillips SM; 'Failed recovery of glycemic control and myofibrillar protein synthesis with two weeks of physical inactivity in overweight, prediabetic older adults'; *J Gerontol A Biol Sci Med Sci.,* Jul 2018
5. Wehrens SMT, Christou S, Isherwood C, Archer SN, Skene DJ and Johnston JD; 'Meal timing regulates the human circadian system'; *Current Biology,* Jun 2017
6. Fowler PM, 'Performance recovery following long-haul international travel in team sport athletes.' *Aspetar Sports Medicine Journal,* Dec 2015
7. www.britishairways.com/travel/drsleep/public
8. www.jetlagrooster.com
9. Schwellnus MP, Derman WE, Jordaan E, Page T, Lambert MI, Readhead C, Roberts C, Kohler R, Collins R, Kara S, Morris MI, Strauss O and Webb S; 'Elite athletes travelling to international destinations >5

time zone differences from their home country have a 2–3-fold increased risk of illness'; *BJSM.*, Sep 2012 *and* Svendsen IS, Taylor IM, Tønnessen E, Bahr R and Gleeson M; 'Training-related and competition-related risk factors for respiratory tract and gastrointestinal infections in elite cross-country skiers'; *BJSM*, Jul 2016

10. Walsh NP; 'Recommendations to maintain immune health in athletes'; *Eur J Sport Sci.*, Jul 2018

11. McFarland LV; 'Meta-analysis of probiotics for the prevention of traveler's diarrhea'; *Travel Med Infect Dis.*, Mar 2007

12. ibid.

13. Young M and Fricker P, 'Medical and nutritional issues for the travelling athlete'; *Clinical Sports Nutrition* (edited by L. Burke and V. Deakin); McGraw Hill, 2006

14. ibid.

15. Fowler PM, 'Performance recovery following long-haul international travel in team sport athletes.' *Aspetar Sports Medicine Journal*, Dec 2015

16. ibid.

17. ibid.

18. Beaumont M, Batéjat D, Piérard C, Van Beers P, Denis JB, Coste O, Doireau P, Chauffard F, French J and Lagarde D; 'Caffeine or melatonin effects on sleep and sleepiness after rapid eastward transmeridian travel'; *J of Applied Physiology,* Jan 2004

Chapter 14. Supplementation: Lifting the Lid

1. Zion Market Research, 'Dietary supplements market by ingredients (botanicals, vitamins, minerals, amino acids, enzymes) for additional supplements, medicinal supplements and sports nutrition applications: Global industry perspective, comprehensive analysis and forecast, 2016–2022'; Jan 2017

2. Huang SH, Johnson K and Pipe AL; 'The use of dietary supplements and medications by Canadian athletes at the Atlanta and Sydney Olympic Games'; *Clin J Sport Med.*, 2006 Jan *and* Sulaiman OA and Salam AI; 'Use of dietary supplements among professional athletes in Saudi Arabia'; *Journal of Nutrition and Metabolism,* Apr 2013

3. Bailey RL, Gahche JJ, Lentino CV, Dwyer JT, Engel JS, Thomas PR, Betz JM, Sempos CT and Picciano MF; 'Dietary supplement use in the United States, 2003-2006'; *J Nutr.*, Feb 2011

4. Paulsen G, Cumming KT, Holden G, Hallen J, Rønnestad BR, Sveen O, Skaug A, Paur I, Bastani NE, Østgaard HN, Buer C, Midttun M, Freuchen F, Wiig H, Ulseth ET, Garthe I, Blomhoff R, Benestad HB and Raastad T; 'Vitamin C and E supplementation hampers cellular adaptation to endurance training in humans:a double-blind, randomised, controlled trial'; *J Physiol.*, Feb 2014

5. Mettler S and Zimmermann MB; 'Iron excess in recreational marathon runners'; *Eur J Clin Nutr.*, May 2010

6. Maughan RJ et al; 'IOC consensus statement: dietary supplements and the high-performance athlete'; *BJSM.*, Apr 2018

7. Brosnan ME and Brosnan JT; 'The role of dietary creatine'; *Amino Acids,* Aug 2016

8. Trexler ET, Smith-Ryan AE, Stout JR, Hoffman JR, Wilborn CD, Sale C, Kreider RB, Jäger R, Earnest CP, Bannock L, Campbell B, Kalman D, Ziegenfuss TN and Antonio J; 'International Society of Sports Nutrition position stand: Beta-alanine'; *J Int Soc Sports Nutr.*, Jul 2015

9. Maughan RJ et al; 'IOC consensus statement: dietary supplements and the high-performance athlete'; *BJSM.*, Apr 2018

10. Jones AM, Thompson C, Wylie LJ and Vanhatalo A; 'Dietary nitrate and physical performance'; *Annu Rev Nutr.*, Aug 2018

11. Jones AM; 'Dietary nitrate supplementation and exercise performance'; *Sports Med.*, May 2014

12. Shaw G, Lee-Barthel A, Ross ML, Wang B and Baar K; 'Vitamin C–enriched gelatin supplementation before intermittent activity augments collagen synthesis'; *The American Journal of Clinical Nutrition,* Jan 2017

13. Maughan RJ et al; 'IOC consensus statement: dietary supplements and the high-performance athlete'; *BJSM.*, Apr 2018

14. Geller AI, Shehab N, Weidle NJ, Lovegrove MC, Wolpert BJ, Timbo BB, Mozersky RP and Budnitz DS; 'Emergency department visits for adverse events related to dietary supplements'; *N Engl J Med.*, Oct 2015

15. Richwine L; 'Hydroxycut products recalled after one death: FDA'; *Reuters,* May 2009

16. Geyer H, Parr MK, Mareck U, Reinhart U, Schrader Y and Schänzer W; 'Analysis of non-hormonal nutritional supplements for anabolic-androgenic steroids: Results of an international study'; *Int J Sports Med.*, 2004

17. Maughan RJ et al; 'IOC consensus statement: Dietary supplements and the high-performance athlete'; *BJSM.*, Apr 2018

18. www.informed-sport.com

19. Mottram DR and Chester N (Eds); *Drugs in Sport*; Oxford Routledge, Feb 2018

Chapter 15. Ageing

1. Jakob J; 'Marathon performance across nations'; April 2018; runrepeat.com/research-marathon-performance-across-nations

2. Figures from Public Health England

3. Figures from Diabetes UK

4. Howden EJ, Sarma S, Lawley JS, Opondo M, Cornwell W, Stoller D, Urey MA, Adams-Huet B and Levine BD;

'Reversing the cardiac effects of sedentary aging in middle age: A randomized controlled trial'; *Circulation,* Jan 2018

5. Hörder H, Johansson L, Guo X, Grimby G, Kern S, Östling S and Skoog I; 'Midlife cardiovascular fitness and dementia: A 44-year longitudinal population study in women'; *Neurology,* Mar 2018

6. Public Health England; 'Physical activity: applying all our health'; Jan 2018; www.gov.uk/government/publications/physical-activity-applying-all-our-health

7. Churchward-Venne TA, Breen L and Phillips SM; 'Alterations in human muscle protein metabolism with aging: Protein and exercise as countermeasures to offset sarcopenia'; *Biofactors,* 2014 Mar–Apr *and* Bauer J, Biolo G, Cederholm T, Cesari M, Cruz-Jentoft AJ, Morley JE, Phillips S, Sieber C, Stehle P, Teta D, Visvanathan R, Volpi E and Boirie Y; 'Evidence-based recommendations for optimal dietary protein intake in older people: a position paper from the PROT-AGE Study Group'; *J Am Med Dir Assoc.,* Aug 2013

8. Smeuninx B, Mckendry J, Wilson D, Martin U and Breen L; 'Age-related anabolic resistance of myofibrillar protein synthesis is exacerbated in obese inactive individuals'; *J Clin Endocrinol Metab.,* Sep 2017

9. www.nhs.uk/live-well/exercise/physical-activity-guidelines-older-adults

10. Phillips SM; 'Current concepts and unresolved questions in dietary protein requirements and supplements in adults'; *Front Nutr.,* May 2017

11. ibid.

12. Oyebode O, Gordon-Dseagu V, Walker A and Mindell JS; 'Fruit and vegetable consumption and all-cause, cancer and CVD mortality: Analysis of health survey for England data'; *J Epidemiol Community Health,* 2014

13. ibid.

14. ibid.

15. Aune D, Giovannucci E, Boffetta P, Fadnes LT, Keum N, Norat T, Greenwood DC, Riboli E, Vatten LJ and Tonstad S; 'Fruit and vegetable intake and the risk of cardiovascular disease, total cancer and all-cause mortality: A systematic review and dose-response meta-analysis of prospective studies'; *International Journal of Epidemiology*, Jun 2017

16. National Statistics; 'Statistics on obesity, physical activity and diet'; March 2017; www.gov.uk/government/statistics/statistics-on-obesity-physical-activity-and-diet-england-2017

17. Feeney M and Reeves L; 'Food fact sheet'; *BDA*, Jul 2017

18. Chumlea WC, Guo SS, Zeller CM, Reo NV and Siervogel RM; 'Total body water data for white adults 18 to 64 years of age: The Fels longitudinal study'; *Kidney International*, Jul 1999

Index

accelerator *see* training/physical exercise
Adele 189
adenosine 94, 185, 193
adenosine triphosphate (ATP) 15, 17–18, 24, 31, 47, 230
ageing 8, 27, 34, 37, 38, 40, 84, 131, 241–57; ageing plan 256–7; ageing process 244, 245; bones and 255–6; hydration and 256; lowering your energy consumption 246–8; muscle maintenance 248–51; protection foods 251–4
alcohol 43, 75, 93, 99–101, 182, 188, 194, 198, 217
amino acids 15, 17, 33, 35, 36, 181, 183, 230, 249, 250
anti-inflammatory effects 32, 181–2, 198, 232–3
antioxidants 14, 34, 40, 84, 93, 162, 181, 196–7, 198, 200, 253, 254
Argonne Diet 219
Arsenal FC 2, 42, 73–4, 134–5, 137, 162, 164, 193

Baar, Professor Keith 233
bacteria, gut 28, 195, 217

base, build your 75
BBC: GoodFood 111, 164, 165; Sport Relief 3, 242–3
behaviours, winning 9, 75, 101, 115–25; eating at home 116, 119–21; eating out 122–5; environment, managing your 116–25; significant others 117–18; training 117; workplace 116–17
Beta-alanine 230–1
Biles, Simone 45
Blanc, Raymond 161
blood glucose 23, 24, 25, 26, 27, 28, 48, 52, 91, 100, 119, 151, 201, 211, 242–3, 246 mood and 156–7
BMI 129–30, 133
body composition, monitoring 128, 131–4
bones 31, 39, 40, 48, 54, 105, 131, 257; ageing process and 244, 245, 255–6; bone mineral density (BMD) 39, 61
Borg scale 138–9
brain 14, 15, 17, 24, 32, 49, 117, 119, 176, 179, 181, 185, 246
break, take a 187
British Athletics 2, 121